PROGRESS ON CRYPTOGRAPHY
25 Years of Cryptography in China

THE KLUWER INTERNATIONAL SERIES
IN ENGINEERING AND COMPUTER SCIENCE

PROGRESS ON CRYPTOGRAPHY
25 Years of Cryptography in China

edited by

Kefei Chen
Shanghai Jiaotong University
China

KLUWER ACADEMIC PUBLISHERS
Boston / Dordrecht / London

Distributors for North, Central and South America:
Kluwer Academic Publishers
101 Philip Drive
Assinippi Park
Norwell, Massachusetts 02061 USA
Telephone (781) 871-6600
Fax (781) 871-6528
E-Mail <kluwer@wkap.com>

Distributors for all other countries:
Kluwer Academic Publishers Group
Post Office Box 322
3300 AH Dordrecht, THE NETHERLANDS
Telephone 31 78 6576 000
Fax 31 78 6576 474
E-Mail <orderdept@wkap.nl>

 Electronic Services <http://www.wkap.nl>

Library of Congress Cataloging-in-Publication

PROGRESS ON CRYPTOGRAPHY
25 Years of Cryptography in China
edited by Kefei Chen

ISBN: 1-4020-7986-9
E-book ISBN: 1-4020-7987-7

International Workshop on Progress on Cryptography

Organized by

Department of Computer Science and Engineering, SJTU

In cooeration with

National Natural Science Foundation of China (NSFC)
Aerospace Information Co., Ltd.

Workshop Co-Chairs

Kefei Chen (Shanghai Jiaotong University, China)
Dake He (Southwest Jiaotong University, China)

Program committee

Kefei Chen (Chair, Shanghai Jiaotong University, China)
Lidong Chen (Motorola Inc., USA)
Cunsheng Ding (HKUST, Hong Kong, China)
Dengguo Feng (Chinese Academy of Sciences, China)
Guang Gong (University of Waterloo, Canada)
Dake He (Southwest Jiaotong University, China)
Xuejia Lai (S.W.I.S. GROUP, Switzerland)
Bazhong Shen, (Broadcom Corp., USA)
Huafei Zhu (Institute for Infocomm Research, Singapore)

Organizing committee

Kefei Chen (Shanghai Jiaotong University, China)
Dawu Gu (Shanghai Jiaotong University, China)
Baoan Guo (Chair, Tsinghua University, China)
Liangsheng He (Chinese Academy of Sciences, China)
Shengli Liu (Shanghai Jiaotong University, China)
Weidong Qiu (Shanghai Jiaotong University, China)
Dong Zheng (Shanghai Jiaotong University, China)

Contents

Foreword   xi

Preface   xiii

Randomness and Discrepancy Transforms   1
Guang Gong

Legendre Sequences and Modified Jacobi Sequences   9
Enjian Bai, Bin Zhang

Resilient Functions with Good Cryptographic Properties   17
WEN Qiao-yan, ZHANG Jie

Differential Factoring for Integers   25
Chuan-Kun Wu

Simple and Efficient Systematic A-codes from Error Correcting Codes   33
Cunsheng Ding, Xiaojian Tian, Xuesong Wang

On Coefficients of Binary Expression of Integer Sums   45
Bao Li, Zongduo Dai

A new publicly verifiable proxy signcryption scheme   53
Zhang Zhang, Qingkuan Dong, Mian Cai

Some New Proxy Signature Schemes from Pairings   59
Fangguo Zhang, Reihaneh Safavi-Naini, Chih-Yin Lin

Construction of Digital Signature Schemes Based on DLP   67
Wei-Zhang Du , Kefei Chen

DLP-based blind signatures and their application in E-Cash systems   73
Weidong Qiu

A Group of Threshold Group-Signature Schemes with Privilege Subsets   81
Chen Weidong, Feng Dengguo

A New Group Signature Scheme with Unlimited Group Size   89
FU Xiaotong, XU Chunxiang

Identity Based Signature Scheme Based on Quadratic Residues   97
Weidong Qiu, Kefei Chen

New Signature Scheme Based on Factoring and Discrete Logarithms   107
Shimin Wei

New Transitive Signature Scheme based on Discreted Logarithm Problem   113
Zichen Li, Juanmei Zhang, Dong Zheng

Blind signature schemes based on GOST signature   123
Zhenjie Huang, Yumin Wang

One-off Blind Public Key   129
Zhang Qiupu, Guo Baoan

Analysis on the two classes of Robust Threshold Key Escrow Schemes   137
Feng Dengguo, Chen Weidong

Privacy-Preserving Approximately Equation Solving over Reals   145
Zhi Gan, Qiang Li, Kefei Chen

An Authenticated Key Agreement Protocol Resistant to DoS attack   151
Lu Haining, Gu Dawu

A comment on a multi-signature scheme   157
ZHENG Dong, CHEN Kefei, HE Liangsheng

Cryptanalysis of LKK Proxy Signature   161
ZHENG Dong, LIU Shengli, CHEN Kefei

Attack on Identity-Based Broadcasting Encryption Schemes   165
Shengli Liu, Zheng Dong, Kefei Chen

Differential-Linear Cryptanalysis of Camellia   173
Wenling WU, Dengguo FENG

Security Analysis of EV-DO System   181
Zhu, Hong Ru

A Remedy of Zhu-Lee-Deng's Public Key Cryptosystem   187
Huafei Zhu, Yongjian Liao

Quantum cryptographic algorithm for classical binary information   195
Nanrun Zhou, Guihua Zeng

Practical Quantum Key Distribution Network   201

Contents

Jie Zhu,Guihua Zeng

A Survey of P2P Network Security Issues based on Protocol Stack   209
ZHANG Dehua, ZHANG Yuqing

DDoS Scouter: A simple IP traceback scheme   217
Chen Kai, Hu Xiaoxin, Hao Ruibing

A Method of Digital Data Transformation–Base91   229
He Dake, He Wei

An approach to the formal analysis of TMN protocol   235
ZHANG Yu-Qing, LIU Xiu-Ying

Foreword

Teacher Xiao will turn 70 this year. As his students, we learnt from him not only scientific knowledge, but also the ethics in the life; not only through the lectures in the serious classroom, but also through the conversations outside the campus over the world, politics, economics, life. We all enjoyed the time of listening your lectures and we are proud to be your students.

Since a quarter of century, teacher Xiao has educated hundreds of us in the fields of mathematics, information theory, communication, cryptology, etc. Today, the "old-classmates" have grown up into the society; many of them are taking the key positions all over the world. Especially, when we talk about the "Xidian branch schools" are spreading the seeds in many places like Beijing, Shanghai, ...

I think he would be proud of the intellect, energy and enthusiasm that he gave us during our campus life and would be especially proud of his achievements and the achievements that his students have made since our Xidian life.

Best wishes to Teacher Xiao's seventieth birthday!

XUEJIA LAI, ZURICH, SWITZERLAND

Preface

This workshop entitled "Progress on Cryptography: 25 Year of Cryptography in China" is being held during the celebration of Professor Guozhen Xiao's 70th birthday. This proceeding is a birthday gift from all of his current and former graduate students, who have had the pleasure of being supervised by Professor Xiao during the last 25 years.

Cryptography, in Chinese, consists of two characters meaning "secret coding". Thanks to Ch'in Chiu-Shao and his successors, the Chinese Remainder Theorem became a cornerstone of public key cryptography. Today, as we observe the constant usage of high-speed computers interconnected via the Internet, we realize that cryptography and its related applications have developed far beyond "secret coding". China, which is rapidly developing in all areas of technology, is also writing a new page of history in cryptography. As more and more Chinese become recognized as leading researchers in a variety of topics in cryptography, it is not surprising that many of them are professor Xiao's former students.

We will never forget a moment in the late 1970's, during the time when China was just opening its door to the world, when Professor Xiao explained the idea of public key cryptography at a lecture. We were so fascinated that many of us have since devoted our careers to cryptography research and applications. Professor Xiao had started a weekly cryptography seminar, where we discussed newly published cryptography research papers from all over the world. We greatly benefited by the method he taught us, which was to catch the main ideas of each piece of research work. He also influenced us deeply by his method of approaching a creative breakthrough. As he said, "only when you can stand on the top of the existing results, just as you stand on the highest peak to look at all the mountains, can you figure out where to go next." With this advice, we took our first step in research by thoroughly understanding other people's work. As a result, many of us generated our first few pieces of work through the seminars.

"Professor Xiao's graduate students" as a group, has been attracting the attention of the academic cryptography community since the first ChinaCrypt in 1984, at which his first few graduate students presented some very impressive work. After 20 years, the research interests of the group have extended to a variety of areas in cryptography. This proceeding includes 32 papers. These papers cover a range of topics, from mathematical results of cryptography to practical applications. This proceeding includes a sample of research conducted by Professor Xiao's former and current graduate students.

In China, we use the term "peaches and plums" to refer to "pupils and disciples". Now Professor Xiao's peaches and plums have spread all over the world. We are recognized as a special group in the cryptography community with not only our distinguished achievements but also our outstanding spirit. Many people have asked about the underlying motivation behind this quarter-century legend in cryptography research, made by professor Xiao and his students. Among all possibilities, I would consider independent thinking and honest attitude as the most crucial aspects. Professor Xiao guided us not only to a fascinating scientific field where many of us made our life-long careers but also to a realm of thought which made us as who we are today.

Please join me in wishing Professor Xiao a Happy 70th Birthday.

LIDONG CHEN, PALATINE, IL, USA

*This proceedings is dedicated
to Professor Guozheng
XIAO on his 70th birthday*

RANDOMNESS AND DISCREPANCY TRANSFORMS

Guang Gong

Department of Electrical and Computer Engineering, University of Waterloo

Waterloo, Ontario N2L 3G1, CANADA

ggong@calliope.uwaterloo.ca

Abstract   In this paper, a new transform of ultimately periodic binary sequences, called a *discrepancy transform*, is introduced in terms of the Berlekamp-Massey algorithm. First, we show that the run property of the discrepancy sequences dominates the randomness of linear span profiles of the sequences. Then, using a modified version of the Berlekamp-Massey algorithm, we provide a method to construct a large family of nonlinear permutations of $GF(2^n)$. Thirdly, applying these permutations as filtering functions to filtering generators, we obtain that the resulting output sequences possess good randomness and have efficient implementations at both hardware and software.

Keywords: discrepancy transform, permutations, filtering generator

1.   Introduction

Pseudo-random sequence generators are widely used in secure communications, such as key stream generators in stream cipher cryptosystems, section key generators in block cipher cryptosystmes, pseudo-random number generators in public-key cryptosystems, and digital watermark.

In 1984, Rueppel [18] addressed the problem that a large linear span can not guarantee unpredictability of a sequence. He then suggested to consider a linear span profile of a sequence as a complement for randomness of the sequence. Since then, a considerable amount of research work has been done along this line [10][11][17]. The linear span profile of a sequence is controlled by runs of zeros in its discrepancy sequence. This allows us to be able to give a definition for smoothly increased linear span profiles in quantity.

By inspiration of the fact that discrepancy sequences dominate the behaviors of linear span profiles, we explore the inverse process for construction of possible good pseudo-random sequence generators. By restricting the discrepancy transform to an n dimensional linear space over $GF(2)$ and using a modified Berlekamp-Massey algorithm, we derive a large family of nonlinear permutations of a finite field $GF(2^n)$, represented in boolean functions. Applying the inverses of these permutations as filtering functions to filter generators, we obtain pseudo-random sequence generators with good randomness, unpredictability, and efficient implementation in both hardware and software.

This paper is organized as follows. In Sections 2 and 3, we introduce the discrepancy transform, and and discuss their application in analysis of randomness of linear span profiles of sequences. In Section 4, we construct a family of permutations of $GF(2^n)$ in terms of a modified Berlekamp-Massey algorithm, and provide randomness properties of a class of filtering generators in which the filtering functions are the inverse discrepancy transforms.

Note. In this paper, we restrict ourselves to $\mathbb{F}_2$. However, all the results obtained here can be easily generalized to an arbitrary finite field. For an introduction of sequence design and analysis, the reader is referred to [4], [18].

2.   Discrepancy Transforms

In this section, we introduce the discrepancy transform and the inverse discrepancy transform. Let us denote $R = \{\{a_i\}|a_i \in \mathbb{F}_2\}$, a ring of binary sequences with infinite elements; $R_0 \subset R$ which contains all ultimately periodic sequences of R; and $R_{-1} = \{(d_0, d_1, \cdots, d_{r-2}, d_{r-1}, 0, 0, \cdots) \in R\} \subset R$, i.e., if $\underline{d} = \{d_i\} \in R_{-1}$, then there is a positive integer such that $d_i = 0, \forall i \geq r$. We denote it as $\{d_i\} = (d_0, d_1, \cdots, d_{r-1}, 0)$, and call r the *ending point* of $\underline{d}$ if $d_{r-1} \neq 0$.

Definition 1. *Let $\underline{a} = \{a_i\} \in R$. For any $n > 0$, and let $LFSR(f_{n-1}, l_{n-1})$ generate a sequence $a_0, a_1, \cdots, a_{n-1}$. We denote $l = l_{n-1}$ and $f_{n-1} = x_l + \sum_{i=0}^{l-1} c_{n-1,i}x^i$, and let $d_n = a_n + \sum_{i=0}^{l-1} c_{n-1,i}a_{n-l+i}, n = 1, 2, \cdots$. Then d_n is called* a next discrepancy bit *of the sequence, and $\{f_i\}$* a linear span profile *of the sequence.*

Definition 2. *Let $\underline{a} = \{a_i\} \in R_0$ be a binary sequence with parameter (u, N), and let $M = 2N + u$. Let $D(\underline{a}) = \{d_n\}$ be a sequence in R_{-1} where d_n is the next discrepancy bit computed by the Berlekamp-Massey Algorithm (BMA, see the Appendix) for $0 \leq n < M$ and $d_n = 0$ for all $n \geq M$. Then D is called* a discrepancy transform *from R_0 to R_{-1}. The sequence $\{d_i\}$ is called a* discrepancy (transform) sequence *of $\{a_i\}$.*

For example, let $\underline{a} = (1001011)$ be a sequence of period 7. Then

$$D(\underline{a}) = \{d_i\} = (1101010000\cdots) = (110101\,\dot{0}).$$

Let $LS(x)$ represents the linear span of a sequence x. Let $\underline{a} = \{a_i\}$ be a sequence in R_0 with period N and $D(\underline{a}) = \{d_i\}$ be the discrepancy sequence of $\underline{a}$. From the BMA, it is clear that if $LS(\underline{a}) = l$, then $d_n = 0, \forall n \geq 2l$.

Theorem 2.1. D *is a bijective map between R_0 and R_{-1}.*

Proof. Let $\underline{a} = \{a_n\} \in R_0$ be an ultimately sequence with parameter (u, N). From the BMA, the polynomials $f_n, 0 \le n < M$, constructed by the BMA are uniquely determined. From the definition of $\{d_n\}$, it is clear that D is an injective. So, it suffices to show that D is surjective. In other words, we need to prove that for any sequence $\underline{d}$ in R_{-1}, there exists an ultimately periodic sequence $\underline{a} \in R_0$ such that $\underline{d}$ is the discrepancy sequence of $\underline{a}$. We can construct a sequence $\underline{a}$ from $\underline{d}$ by switching the places of a_n and d_n in the BMA (the details are omitted here due to short of space). Therefore LFSR (f_{r-1}, l_{r-1}) generates the sequence $\underline{a}$. Thus $\underline{a}$ is an ultimately periodic sequence with the parameter (u, t) where $t = per(g)$ and $f_{r-1} = x^t g(x)$ with $g(0) \ne 0$ (see [13]). In other words, we get that $a_{n+t} = a_n$ for all $n \ge u$. So, $\underline{a} \in R_0$ and $D(\underline{a}) = \underline{d}$. Thus D is a surjective map from R_0 to R_{-1}. Therefore D is a bijective map between R_0 and R_{-1}. □

According to Theorem 2.1, D is invertible and $D^{-1}(\underline{d})$ can be constructed by the proof of Theorem 2.1. The inverse map D^{-1} of D is called the *inverse discrepancy transform (IDT)*, and the sequence $D^{-1}(\underline{d})$ an *inverse discrepancy (transform) sequence* of $\underline{d}$. From the proof of Theorem 2.1, we have the following result on the inverse discrepancy sequences.

Corollary 1. *With the notation in Theorem 2.1.*

(a) f_{r-1} *is the minimal polynomial of $D^{-1}(\underline{d})$, so that l_{r-1} is the linear span of the inverse discrepancy sequence, i.e., $LS(D^{-1}(\underline{d})) = l_{r-1}$. Furthermore, $l_{r-1} = \lceil r/2 \rceil$ where $\lceil x \rceil$ represents the least integer that is not less than x.*

(b) $D^{-1}(\underline{d})$ *is an ultimately periodic sequence with the parameter (u, t) where $u \le l_{r-1}$ and $t = per(g)$ where $f_{r-1}(x) = x^t g(x)$ with $g(0) \ne 0$.*

Example 1. Let $\underline{d} = (1001011\ \dot{0}) \in R_{-1}$ with $r = 7$. Then

$$D^{-1}(\underline{d}) = \underline{a} = 1110111101 \cdots \in R_0$$

which is a periodic sequence with period 5, i.e, $a_{n+5} = a_n$ for all $n \ge 0$. Furthermore, $f_6 = x^4 + x^3 + x^2 + x + 1$ has period 5. Note that the first 7 elements of $\underline{d}$ are taken from the elements in a period of an m-sequence with period 7.

Example 2. Let $\underline{d} = (110011010001011\ \dot{0}) \in R_{-1}$ with $r = 15$. Then

$$D^{-1}(\underline{d}) = 100011100010101101000011001001111\underline{101110} \cdots .$$

Hence $f_{14}(x) = x^8 + x^7 + x^6 + x^5 + x^3 = x^3 g(x)$ where $g(x) = x^5 + x^4 + x^3 + x^2 + 1$ a primitive polynomial over $\mathbb{F}_2$. Therefore $LS(D^{-1}(\underline{d})) = 8$ and

$D^{-1}(\underline{d}) = \underline{a}$ is an ultimately periodic sequence with the parameter $(3, 31)$, i.e., $a_{n+31} = a_n$ for all $n \geq 3$. Note that the first 15 element of $\underline{d}$ are taken from the elements of a period of a modified de Bruijn sequence [15] with period 15.

3.   Runs of Discrepancy Sequences and Linear Span Profiles

In this section, we first show the randomness of the linear span profile of a sequence is dominated by its discrepancy transform sequence. We then give a criterion for a smoothly increased linear span profile and an optimal linear span by means of runs of the discrepancy sequence. By carefully determining the values of a and b in the Berlekamp-Massey algorithm, we can establish the following results (the proof will be provided in the full paper).

Theorem 3.1. *Let* $\underline{a} \in R_0$ *and* $\underline{d}$ *be its discrepancy sequence. Let* $\mu(\underline{d}^r)$ *be the greatest length of runs of 0's in* $\underline{d}^r = (d_0, d_1, \cdots, d_{r-1})$. *Then* $\{l_n\}$, *the linear span profile of* $\underline{a}$, *satisfies* $l_n - l_{n-1} \leq \mu(\underline{d}^r) + 1$, $\forall n > 0$.

Corollary 2. *With the notation in Theorem 3.1. For any* $n > 0$, $l_n - l_{n-1} = a + 1$ *where* $a = n - 1 - j$ *where* j *is the largest number in a set of* $\{m, m+1, \cdots, n-1\}$ *such that* $d_j = 1$ *and* $(d_{j+1}, d_{j+2}, \cdots, d_{n-1})$ *is a run of 0's where* m *is an integer satisfying* $l_{m-1} < l_m = l_{m+1} = \cdots = l_{n-1}$. *In other words, the difference between* l_n *and* l_{n-1} *is equal to the length of the run of 0's preceded to* d_n *plus one.*

According to Theorem 3.1 and Corollary 2, the behavior of the linear span profile of a periodic sequence is completely determined by lengths of runs in the discrepancy sequence. More precisely, given a sequence $\underline{d} \in R_{-1}$, a pseudo-random sequence generator (PSG) generates an inverse discrepancy sequence $D^{-1}(\underline{d}) = \{a_i\}$ in the following fashion. At each clock cycle $n > 0$, if $d_n = 0$, then the PSG uses the previous LFSR to generate a current bit a_n. If $d_n = 1$, then the PSG reloads a new LFSR to generate a current bit a_n. So the nth bit of output of the PSG is generated by the previous LFSR or a new LFSR depending on $\{d_n\}$. In the discrepancy sequence, a run of 0's of length t means that the PSG does not change the LFSR during t consecutive clock cycles. A run of 1's of length t means that the PSG changes LFSR at each clock cycle during t consecutive clock cycles where the lengths of these LFSRs may not change. The randomness of runs of a sequence is given by the Golomb Randomness Postulate R-2. If the discrepancy sequence satisfies the randomness postulate R-2, then the frequency that the PSG changes LFSRs can be considered as a random variable with a uniform distribution. We summarize these discussions into the following criteria for measuring randomness of pseudo-random sequences.

Let $\underline{a}$ be a sequence of period N and $\underline{d} = (d_0, d_1, \cdots, d_{r-1}, 0)$ be its discrepancy sequence. Note that if a sequence of period N or length N satisfies the randomness postulate R-2, then the greatest length of runs in the sequence is bounded by $\sigma(N) = \lceil log_2^N \rceil + 1$. So $\sigma(r)$ is the best bound for $\mu(\underline{\mathbf{d}}^r)$, the largest length of the runs of zeros in $\underline{\mathbf{d}}^r$.

Randomness Criteria of Linear Spans: (a) If $\mu(\underline{\mathbf{d}}^r) < \sigma(r)$ for any shift of $\underline{\mathbf{a}}$, then we say that $\underline{\mathbf{a}}$ has a *smoothly increased linear span profile.* (b) If $\underline{\mathbf{d}}^r$ satisfies the randomness postulate R-2 for any shift of $\underline{\mathbf{a}}$ and $LS(\underline{\mathbf{a}})$, the linear span of $\underline{\mathbf{a}}$, satisfies that

$$N/2 - \epsilon < LS(\underline{\mathbf{a}}) < N \text{ where } 0 \le \epsilon \le c\sigma(r) \tag{1}$$

where $c \ge 0$ is a constant, then we say that $\underline{\mathbf{a}}$ has an *optimal linear span.*

We tested some known generators with small parameters. For example, we considered three types of known pseudo-random sequences whose linear spans satisfy (1), i.e., de Bruijn sequences [3] with period 2^n, the self-shrink sequences [16] with period 2^{n-1}, and the elliptic curve sequences of type I [6] with period 2^{n+1} where n is the parameter related to their respective constructions. If $2^n - 1$ is a prime, then we have quadratic sequences with period $2^n - 1$. For their discrepancy sequences, none of them satisfies the randomness postulate R-2. However, the experimental results showed that some of them did satisfy the condition for smoothly increased linear span profiles.

When we use the inverse process to generate pseudo-random sequences discussed above, it is clear that the nth bit depends on the previous $n - 1$ bits. Thus it is impossible to hold or store the entire bits of an inverse discrepancy sequence in practical cryptosystems. How to generate a sequence while considerably preserving the features provided by the inverse discrepancy sequences with good randomness and considerably reduced the computational cost in both time and space is the purpose of the remaining section.

4. Restricted Discrepancy Transforms and Filtering Generators with D-Permutations

In this section, we first discuss a restriction of the discrepancy transform on $\mathbb{F}_2^{(n)} = \{(x_0, \cdots, x_{n-1}) \mid x_i \in \mathbb{F}_2\}$ and how to construct a large family of permutations resulted from the restricted discrepancy transform. We then present randomness properties of filtering generators in which the filtering functions are the nth component of the constructed permutations. Let $V = \mathbb{F}_2^{(n)}$. Then V can be embedded into R_0 [8] via

$$(x_0, x_1, \cdots, x_{n-1}) \longmapsto (x_0, x_1, \cdots, x_{n-1}, x_0, x_1, \cdots, x_{n-1}, \cdots) \in R_0.$$

Thus we have D_V, a restriction of D on V, as follows

$$D_V(a_0, a_1, \cdots, a_{n-1}) = (d_0, d_1, \cdots, d_{n-1}) \tag{2}$$

where $d_i, 0 \leq i < n$ are computed by the BMA. Note that any function from V to V can be represented by it's n component functions. In other words, we can write

$$D_V(x_0, x_1, \cdots, x_{n-1}) = (g_{n,0}, g_{n,1}, \cdots, g_{n,n-1}), x_i \in \mathbb{F}_2$$

where $g_{n,j} = g_{n,j}(x_0, x_1, \cdots, x_{n-1})$ is a function from V to $\mathbb{F}_2$, i.e., a Boolean function in n variables $x_0, x_1, \cdots, x_{n-1}$.

Lemma 1. *D_V is a permutation of $\mathbb{F}_{2^n}$.*

Proof. According to Theorem 2.1, D_V is a bijective map on V. Since V is isomorphic to the finite field $\mathbb{F}_{2^n}$, then D_V is a permutation of $V = \mathbb{F}_{2^n}$. $\square$

We call D_V a *restricted discrepancy transform* on V and D_V^{-1} the *inverse restricted discrepancy transform* on V.

Theorem 4.1. *Let $D_V = (g_{n,0}, g_{n,1}, \cdots, g_{n,n-1})$ be the restricted discrepancy transform on V. Then D_V is an nonlinear permutation of $\mathbb{F}_{2^n}$ for $n > 2$ for which*

$$g_{n,i}(x_0, x_1, \cdots, x_{n-1}) = g_{i+1,i}(x_0, x_1, \cdots, x_i), 0 \leq i < n \qquad (3)$$

Precisely, for $i = 0, 1, 2,$ and 3, we have $g_{1,0}(x_0) = x_0$, $g_{2,1}(x_0, x_1) = x_0 + x_1$, $g_{3,2}(x_0, x_1, x_2) = x_0 x_1 + x_2$, and $g_{4,3}(x_0, x_1, x_2, x_3) = x_1 + x_1 x_2 + x_0 x_1 x_2 + x_3$, and

$$g_{n,n-1}(x_0, x_1, \cdots, x_{n-1}) = h(x_0, x_1, \cdots, x_{n-2}) + x_{n-1}, n > 1 \qquad (4)$$

where $h(x_0, x_1, \cdots, x_{n-2})$ is a Boolean function in $n - 1$ variables.

A proof of this result will be provided in the full version of this work. The inverse restricted discrepancy transform D_V^{-1} has similar properties as those of D_V.

Corollary 3. *Let $D_V^{-1}(x)(f_{n,0}, f_{n,1}, \cdots, f_{n,n-1})$ be the inverse of D_V. Then D_V^{-1} is nonlinear for $n > 2$ and*

$$D_V^{-1}(x_0, x_1, \cdots, x_{n-1}) = (f_{1,0}, f_{2,1}, f_{3,2}, f_{4,3}, \cdots, f_{n,n-1})$$

where

$$f_{i,i-1} = f_{i,i-1}(x_0, x_1, \cdots, x_{i-1}), 1 \leq i \leq n.$$

Precisely, for $i = 1, 2, 3,$ and 4, we have $f_{1,0}(x_0) = x_0 = g_{1,0}$, $f_{2,1}(x_0, x_1) = x_0 + x_1 = g_{2,1}$, $f_{3,2}(x_0, x_1, x_2) = x_0 + x_0 x_1 + x_2$, and $f_{4,3}(x_0, x_1, x_2, x_3) = x_0 + x_1 + x_1 x_2 + x_0 x_1 x_2 + x_3$.

By this method, for fixed n, we can construct only one pair of nonlinear permutations D_V and D_V^{-1} from the BMA. In order to construct a family of permutations on V in terms of the discrepancy transform, we modify the initial step and the loop step in the BMA (see the Appendix) as follows. For $0 \leq n_0 < n$ and $0 < v < n$, let

$$R = \{x^{n_0+1} + \sum_{i=1}^{n_0} t_i x^i + 1 | t_i \in \mathbb{F}_2\}, \tag{5}$$

and

$$U = \{x^v + \sum_{i=0}^{v-1} k_i x^i \mid k_i \in \mathbb{F}_2\}. \tag{6}$$

At the initial step, choose one of polynomials in R, say $r_{n_0}(x)$, to generate the sequence $(a_0, a_1, \cdots, a_{n_0}) = (0, 0, \cdots, 0, 1)$. At the loop step, if $d_n = 1$, we select one of polynomials in U, say $u_v(x)$. The rest of the procedure remains. In this way, we can construct at least $2^{n(n-2)/4}$ if n even, and $2^{(n-1)(n-3))/4}$ if n odd permutations of $GF(2^n)$.

In the following, we present the randomness properties of filtering generators for which the filtering functions are inverse D-permutations. Let $g(x)$ be a D-permutation on V. We can write $g^{-1}(x)$, the inverse of $g(x)$, as follows

$$g^{-1}(x) = (g_{1,0}^{-1}, g_{2,1}^{-1}, \cdots, g_{n,n-1}^{-1}), x \in V.$$

Let $f = g_{n,n-1}^{-1}(x_0, x_1, \cdots, x_{n-1})$ which is the nth component function of the D-permutation $g^{-1}(x)$, $\{d_i\}$ a binary m-sequence of degree n, and

$$a_i = f(d_i, d_{i+1}, \cdots, d_{i+n-1}), i = 0, 1, \cdots. \tag{7}$$

Then we say that the sequence $\{a_i\}$ is a *D-filter sequence* and f a *D-filter function*.

Randomness profile for D-filter sequences: Any D filter sequence has period $2^n - 1$ and is balanced. Furthermore, all D-filter sequences are shift-distinct. Precisely, there are $\phi(2^n - 1)/n$ shift distinct D-filter sequences with D-filter function f.

The experimental results show that most of shift-distinct D-filter sequences achieve the maximal linear span $2^n - 2$ for every f and a few of them have the linear spans taken on the slightly smaller value $2^n - 2 - (2n + m)$ where $m|n$ or $m = 0$. Therefore, we have the following conjecture for linear spans of the D-filtering sequences.

Conjecture. The linear span of $\{a_i\}$ is equal to $2^n - 2$ or $2^n - 2 - (2n+m)$ where $m|n$ or $m = 0$.

The validity of the conjecture was verified for $4 \leq n \leq 13$.

5. Conclusion

In terms of the Berlekamp-Messay algorithm, we introduced the discrepancy transform for ultimately periodic sequences. Randomness criteria for linear span profiles of sequences are obtained in terms of runs of discrepancy transform sequences. A restriction of the discrepancy transform, computed by the modified Berlekamp-Messay algorithm, derives a new family of nonlinear permutations of $GF(2^n)$. Applying the nth component function of such a permutation to a filter generator yields a pseudorandom sequence generator with strong cryptographic properties, which have potential applications in secure communications.

References

[1] Berlekamp, E.R.,*Algebraic coding theory*, McGraw-Hill, New York, 1968.

[2] de Bruijn, N.G., A combinatorial problem, *Kononklijke Nederlands Akademi van Wetenchappen, Proc.*, vol. 49, Pr. 2, 1946.

[3] Chan, A.H., et al., On the complexities of de Bruijn sequences, *J. Combin. Theory*, vol. 33, Nov. 1982.

[4] Golomb, S.W. (1982) *Shift Register Sequences*, Revised Edition, Aegean Park Press.

[5] Gong, G., On q–ary cascaded GMW sequences, *IEEE Trans.*, IT-42, No. 1, 1996.

[6] Gong, G., et al., Elliptic curve pseudo-random sequence generators, *Proc. of the Sixth Annual Workshop on Selected Area in Cryptography*, August 9-10, 1999, Kingston, Canada.

[7] Herlestam, Tor, On functions of linear shift register sequences, *EuroCrypt'85*, LNCS 219, Springer-Verlag, 1985.

[8] Jacobson, N. (1974) *Basic Algebra I*, W.H. Freeman and Company, San Francisco.

[9] Key, E.L., An analysis of the structure and complexity of nonlinear binary sequence generators, *IEEE Trans.*, IT-22, No. 6, 1976.

[10] Niederreiter, H., Keystream sequences with a good linear complexity profile for every starting point, *EUROCRYPT'89*, LNCS 434, Springer-Verlag, Berlin, 1990.

[11] Niederreiter, H., Some computable complexity measures for binary sequences, *Proc. of SETA'98, Discrete Math. and Theoretical Computer Sci.*, Springer-Verlag, Berlin, 1999.

[12] Niederreiter, H. et al., Simultaneous shifted continued fraction expansions in quadratic time, *Applicable Algebra Engrg. Comm. Comput.* **9** (1998).

[13] Lidl, R. et al., *Finite Fields*, Encyclopedia of Mathematics and its Applications, Volume 20, Addison-Wesley, 2001(revised version).

[14] Massey, J.L., Shift-register synthesis and BCH decoding, *IEEE Trans.*, IT-15, 1969.

[15] Mayhew, G.L. et al., Linear spans of modified de Bruijn sequences, *IEEE Trans.*, IT-36, No. 5, 1990.

[16] Meier, W.,et al., The self-shrinking generator, *EUROCRYPT'94*, LNCS 950, Springer-Verlag, Berlin, 1995.

[17] Paper, F., Stream cipers, *Electrotechnik und Maschinenbau* **104** (1987).

[18] Rueppel, R.A., *Analysis and Design of Stream Ciphers*, Springe-Verlag, 1986.

[19] Welch, L.R. et al., Continued fractions and Berlekamp's algorithm, *IEEE Trans.*, IT-25, 1979.

LEGENDRE SEQUENCES AND MODIFIED JACOBI SEQUENCES

Enjian Bai

Key Lab. on ISN, Xidian University, Xi'an, 710071, China
ejbai@sohu.com

Bin Zhang

Key Lab. of Inform. Security, Graduate School of Chinese Academy of Sciences, Beijing, China
mzb_123@hotmail.com

Abstract   In this paper, a survey of Legendre sequences and modified Jacobi sequences is presented, firstly. We introduce the construction and periodic autocorrelation functions of these two sequences (binary and polyphase). Then we determine the linear complexity of all modified polyphase Jacobi sequences and the corresponding feedback polynomials of the shortest linear feedback shift register that generates such a sequence. Making use of these results, at the same time, we prove the conjectures on the linear complexity and feedback polynomials of modified Jacobi sequences brought forward by D.H. Green and J. Choi.

Keywords: Legendre sequence, modified Jacobi sequence, modified polyphase Jacobi sequence, linear complexity, periodic autocorrelation functions

Introduction

Pseudorandom sequences with good periodic or aperiodic autocorrelation properties are extremely useful in many areas such as communication and cryptography [1, 2]. For cryptographic applications of sequences as key stream ciphers, their linear complexity c_L, is an important figure-of-merit. Legendre sequences (Polyphase Legendre sequences), *L-sequences (PL-sequences) for short*, and modified Jacobi sequences (modifies polyphase Jacobi sequences), *MJ-sequences (MPJ-sequences) for short*, possess good periodic correlation properties and have high linear complexity, which give them some cryptographic significance [3, 4, 5, 6, 7, 8, 9].

This paper will investigate the construction and properties of these two sequences firstly, and then determine the linear complexity and feedback polynomials of MPJ-sequences. At the same time, we prove the conjecturs on the linear complexity and feedback polynomials of MJ-sequences brought forward by D.H. Green and J. Choi [7].

1.   Legendre sequences

Binary Legendre or quadratic residue sequences exist for all lengths L which are prime. They can be constructed using the Legendre symbol (i/L), a Legendre sequence $\hat{a}_0\hat{a}_1\hat{a}_2\cdots\hat{a}_{L-1}$ is then formed by writing $\hat{a}_i = (i/L)$ $0 < i < L$ and the value of $\hat{a}_0$ can be taken either as 1 or -1. Alteratively, a pure binary form of these sequences $a_0a_1a_2\cdots a_{L-1}$, with $a_i = 0$ or 1, can be constructed by mapping the square roots of unity onto the binary symbols in the normal way, i.e., $1 = -1^0$, $-1 = -1^1$, so $1 \to 0$, $-1 \to 1$. Thus this is equivalent to taking

$$a_i = \left(\frac{i}{p}\right) = \begin{cases} 0 & \text{if } i \text{ is a quadratic residue mod } p \\ 1 & \text{if } i \text{ is a quadratic nonresidue mod } p \end{cases} \tag{1}$$

This gives rise to two classes of L-sequences.

Class 1: $L \equiv 3 \bmod 4$. The periodic autocorrelation function $R(\tau)$ takes values $R(\tau) \in \{L, -1\}$ and so this class has the ideal two-valued autocorrelation function. The sequence conventionally referred as quadratic residue sequences belong to this class.

Class 2: $L \equiv 1 \bmod 4$. For this case $R(\tau) \in \{L, -3, 1\}$ and so this class has a three-valued autocorrelation function.

L-sequences have a number of interesting properties [3, 4], C. Ding and T. Helleseth determined the linear complexity of all L-sequences and their minimal polynomials in [5]. These results can be summarized as follows:

$$c_L = \begin{cases} (L+1)/2 & \text{if } L \equiv 1 \bmod 8 \\ L-1 & \text{if } L \equiv 3 \bmod 8 \\ L & \text{if } L \equiv 5 \bmod 8 \\ (L-1)/2 & \text{if } L \equiv 7 \bmod 8 \end{cases} \tag{2}$$

$$m(x) = \begin{cases} (x+1)\cdot f(x) & \text{if } L \equiv 1 \bmod 8 \\ (x^L+1)/(x+1) & \text{if } L \equiv 3 \bmod 8 \\ (x^L+1) & \text{if } L \equiv 5 \bmod 8 \\ f(x) & \text{if } L \equiv 7 \bmod 8 \end{cases} \quad \text{over GF(2)} \tag{3}$$

where $f(x)$ is a special polynomial of degree $(L-1)/2$, that is derived from the sequence.

2.   Modified Jacobi sequences

Firstly, we introduce the Jacobi sequences, which constructed by combining two L-sequences. Jacobi sequences exist for all lengths of the form $L = uv$, where u and v are both prime. They are constructed using the Jacobi symbol $[i/uv]$, which is defined as

$$\left[\frac{i}{uv}\right] = \left(\frac{i}{u}\right) \oplus \left(\frac{i}{v}\right) \qquad 0 \le i < L \tag{4}$$

the term-by-term modulo 2 addition in the 0, 1 form. A Jacobi sequence $b_0 b_1 b_2 \cdots b_{L-1}$ is then formed by writing

$$b_i = \left[\frac{i}{L}\right] = \left(\frac{i}{u}\right) \oplus \left(\frac{i}{v}\right) \qquad 0 < i < L \tag{5}$$

The Jacobi sequences described above do not show particularly good auto-correlation functions and contain out-of-phase values which are related to the factor u and v.

If a Jacobi sequence is modified by ensuring that $b_i = 0$ for $i \equiv 0 \bmod v$, and $b_i = 1$ for $i \ne 0$ and $i \equiv 0 \bmod u$, the resulting sequence called modified Jacobi sequence has greatly improved periodic autocorrelation values [6]. It is assumed, without loss of generality, that $v > u$, so that $v = u + k$, where k is an even integer. If $k \equiv 2 \bmod 4$, the autocorrelation values are taken from $\{L, k - 3, -1, 1 - k\}$, and if $k \equiv 0 \bmod 4$, they are taken from $\{L, k - 3, 1, -3, 1 - k\}$.

D.H. Green and J. Choi conjectured the linear complexity and feedback polynomials of MJ-sequences [7]. We will prove their conjectures in the follow section.

3. Polyphase Legendre sequences

PL-sequences were called polyphase power residue sequences in [8]. Let $a_0 a_1 a_2 \cdots a_{L-1}$ be a q-phase L-sequence of length L, where both L and q are prime such that $L \equiv 1 \bmod 2q$. Let $T = (L - 1)/q$ and μ be a primitive element $\bmod L$, then each non-zero integer $i \bmod L$ can be represented as $i = \mu^t \bmod L \in C_k = \mu^k \{(\mu^0)^q, (\mu^1)^q, \cdots, (\mu^{T-1})^q\} \bmod L, k = 0, 1, \cdots, q-1$. Then, make

$$a_i = j \text{ if } i \in C_j \quad \text{for } 0 < i \le L - 1 \tag{6}$$

and a_0 can be selected to be any of the q available values. We assume, unless otherwise stated, that $a_0 = 0$.

The linear complexity of these sequences has been derived and revealed that it depends on whether q is a qth power residue and the value chosen for the initial digit in the sequence. These results can be summarized as follows:

$$c_L = \begin{cases} L & \text{if } b_0 \ne 0 \text{ and } p \notin C_0 \\ L - 1 & \text{if } b_0 = 0 \text{ and } p \notin C_0 \\ L - T & \text{if } b_0 \ne 0 \text{ and } p \in C_0 \\ L - T - 1 & \text{if } b_0 = 0 \text{ and } p \in C_0 \end{cases} \tag{7}$$

$$m(x) = \begin{cases} x^L - 1 & \text{if } b_0 \neq 0 \text{ and } p \notin C_0 \\ (x^L - 1)/(x - 1) & \text{if } b_0 = 0 \text{ and } p \notin C_0 \\ \frac{x^L - 1}{\omega_i(x)} & \text{if } b_0 \neq 0 \text{ and } p \in C_0 \\ \frac{x^L - 1}{(x-1) \cdot \omega_i(x)} & \text{if } b_0 = 0 \text{ and } p \in C_0 \end{cases} \qquad (8)$$

where $\omega_i(x)$ is the polynomial corresponding to the coset C_i which provide the roots of $B(x) = b_0 + b_1 x + \cdots + b_{L-1} x^{L-1}$.

4.   Modified polyphase Jacobi sequences

Let $A = a_0 a_1 \cdots a_{u-1}$ and $B = b_0 b_1 \cdots b_{v-1}$ be two q-phase L-sequences of length u and v, respectively, where u, v, q are both odd prime and such that $u \equiv 1 \bmod 2q$, $v \equiv 1 \bmod 2q$. Define sequence $R = r_0 r_1 \cdots r_{L-1}$ of length $L = uv$ as $r_i = (a_i + b_i) \bmod q, 0 \leq i < L$.

Sequences with a length L which can be factorized into two or more relatively prime factors can be folded into a two-dimensional structure sometimes referred to as pesudorandom array (PRA) [10]. One method for performing this folding is to start at the top left-hand corner of the array with the first digit of the sequence, and then to place subsequent digits down the diagonal by moving one position in each dimension at each step. When an edge is encountered, the array is re-entered at the opposite edge on the next row or column. In this way, each location in the array will be visited exactly once if one pass through the sequence, provided the dimensions of the array are relatively prime.

The MPJ-sequence $S = s_0 s_1 \cdots s_{L-1}$ of length $L = uv$ is defined as:

$$s_i = \begin{cases} 0 & i = 0 \\ r_i & 0 < i \leq L - 1, (i, L) = 1 \\ m & i \equiv 0 \bmod u \\ n & i \equiv 0 \bmod v \end{cases} \qquad (9)$$

where $0 \leq m \neq n \leq q - 1$. Here we restrict that $n = 0, m \neq 0$.

From the definition above, a MPJ-sequence S can be represented as a $u \times v$ array and it can be decomposed as a modulo-q sum of four component arrays. Then S can be thought of as a modulo-q sum of the following four component sequences of length uv

$$\begin{array}{l} a_0 a_1 \cdots a_{u-1} a_0 a_1 \cdots a_{u-1} \cdots a_0 a_1 \cdots a_{u-1} \\ b_0 b_1 \cdots b_{v-1} b_0 b_1 \cdots b_{v-1} \cdots b_0 b_1 \cdots b_{v-1} \\ 0 \underbrace{0 \cdots 0}_{v-1} (-a_v) \underbrace{0 \cdots 0}_{v-1} (-a_{2v}) \cdots (-a_{(u-1)v}) \underbrace{0 \cdots 0}_{v-1} \\ 0 \underbrace{0 \cdots 0}_{u-1} (m - b_u) \underbrace{0 \cdots 0}_{u-1} (m - b_{2u}) \cdots (m - b_{(v-1)u}) \underbrace{0 \cdots 0}_{u-1} \end{array} \qquad (10)$$

if these sequences are unfolded from the array.

Let $S(x) = s_0 + s_1 x + \cdots + s_{L-1} x^{L-1}$. It is easy to see that [11]
(1) The feedback polynomial $m(x)$ of S is given by

$$(x^L - 1)/\gcd(x^L - 1, S(x)). \tag{11}$$

(2) The linear complexity c_L of S is given by

$$c_L = L - \deg[\gcd(x^L - 1, S(x))]. \tag{12}$$

Since $(L, q) = 1$, there exists a primitive Lth root of unity α in some splitting field $\mathrm{GF}(q^n)$ of $x^L - 1$, and $\gcd(x^L - 1, S(x))$ will be given by the number of values for j, where $0 \le j < L$, such that $S(\alpha^j) = 0$, hence

$$c_L = L - |\{j : S(\alpha^j) = 0, 0 \le j \le L - 1\}|. \tag{13}$$

From (10), the authors can write

$$
\begin{aligned}
S(x) &= \sum_{i=0}^{L-1} s_i x^i \\
&= \left(\sum_{i=0}^{u-1} a_i x^i + \sum_{i=0}^{u-1} a_i x^{i+u} + \cdots + \sum_{i=0}^{u-1} a_i x^{i+(v-1)u} \right) \\
&\quad + \left(\sum_{i=0}^{v-1} b_i x^i + \sum_{i=0}^{v-1} b_i x^{i+v} + \cdots + \sum_{i=0}^{v-1} b_i x^{i+(u-1)v} \right) \\
&\quad + \left(\sum_{i=1}^{u-1} (-a_{iv}) x^{iv} \right) + \left(\sum_{i=1}^{v-1} (m - b_{iu}) x^{iu} \right) \\
&= \left(\sum_{i=0}^{v-1} x^{ui} \right) \left(\sum_{i=1}^{u-1} a_i x^i \right) + \left(\sum_{i=0}^{u-1} x^{vi} \right) \left(\sum_{i=1}^{v-1} b_i x^i \right) \\
&\quad - \sum_{i=1}^{u-1} a_{vi} x^{vi} - \sum_{i=1}^{v-1} b_{ui} x^{ui} + m \sum_{i=1}^{v-1} x^{ui} \\
&= \left(\sum_{i=0}^{v-1} x^{ui} \right) \left(0 \sum_{i \in C_0^u} x^i + \cdots + (q-1) \sum_{i \in C_{q-1}^u} x^i \right) \\
&\quad + \left(\sum_{i=0}^{u-1} x^{vi} \right) \left(0 \sum_{i \in C_0^v} x^i + \cdots + (q-1) \sum_{i \in C_{q-1}^v} x^i \right) \\
&\quad - \left(0 \sum_{vi \in C_0^u} x^{vi} + \cdots + (q-1) \sum_{vi \in C_{q-1}^u} x^{vi} \right)
\end{aligned}
$$

$$-\left(0\sum_{ui\in C_0^v} x^{ui} + \cdots + (q-1)\sum_{ui\in C_{q-1}^v} x^{ui}\right) + m\sum_{i=1}^{v-1} x^{ui}$$

So, for $x = \alpha^0 = 1$,

$$\begin{aligned} S(1) &= v\cdot\frac{(u-1)(q-1)}{2} + u\cdot\frac{(v-1)(q-1)}{2} - \frac{(u-1)(q-1)}{2} \\ &\quad -\frac{(v-1)(q-1)}{2} + m(v-1) \equiv 0 \bmod q \end{aligned} \tag{14}$$

For $x = \alpha^j$, $j \in \Delta_1 = \{j : j \equiv 0 \bmod u, j \neq 0\}$,

$$\begin{aligned} S(\alpha^j) &= u\left(0\sum_{i\in C_0^v}(\alpha^j)^i + \cdots + (q-1)\sum_{i\in C_{q-1}^v}(\alpha^j)^i\right) \\ &\quad -\left(0\sum_{i\in C_0^v}(\alpha^j)^i + \cdots + (q-1)\sum_{i\in C_{q-1}^v}(\alpha^j)^i\right) \\ &\quad -\frac{(u-1)(q-1)}{2} + m(-1) \equiv -m \bmod q. \end{aligned} \tag{15}$$

For $x = \alpha^j$, $j \in \Delta_2 = \{j : j \equiv 0 \bmod v, j \neq 0\}$,

$$\begin{aligned} S(\alpha^j) &= v\left(0\sum_{i\in C_0^u}(\alpha^j)^i + \cdots + (q-1)\sum_{i\in C_{q-1}^u}(\alpha^j)^i\right) \\ &\quad -\left(0\sum_{i\in C_0^u}(\alpha^j)^i + \cdots + (q-1)\sum_{i\in C_{q-1}^u}(\alpha^j)^i\right) \\ &\quad -\frac{(v-1)(q-1)}{2} + m(v-1) \equiv 0 \bmod q. \end{aligned} \tag{16}$$

For $x = \alpha^j$, $j \in \Delta_3 = Z_L - \Delta_1 - \Delta_2 - \{0\}$,

$$\begin{aligned} S(\alpha^j) &= -\left(0\sum_{vi\in C_0^u}(\alpha^j)^{vi} + \cdots + (q-1)\sum_{vi\in C_{q-1}^u}(\alpha^j)^{vi}\right) \\ &\quad -\left(0\sum_{ui\in C_0^v}(\alpha^j)^{ui} + \cdots + (q-1)\sum_{ui\in C_{q-1}^v}(\alpha^j)^{ui}\right) \\ &\quad -m \bmod q. \end{aligned} \tag{17}$$

Then by (14–19), we have

$$c_L = L - u - |\{j : S(\alpha^j) = 0, j \in \Delta_3\}|. \tag{18}$$

When $j \in \Delta_3$, we have the following basic fact:

Fact 1 *When* $j \in \Delta_3$, $S(\alpha^j) \in \{0, 1, \cdots, q-1\}$ *if and only if* $q \in C_k^u$, $q \in C_{q-k}^v$, $k = 0, 1, \cdots, q-1$.

The proof of the fact can be found in our another submission. Then when $q \in C_k^u, q \in C_{q-k}^v, |\{j : S(\alpha^j) = 0, j \in \Delta_3\}| = (L - u - v - 1)/q$. In other cases, $|\{j : S(\alpha^j) = 0, j \in \Delta_3\}| = 0$. So the linear complexity and feedback polynomials of all MPJ-sequences can be determined as follows:

$$c_L = \begin{cases} qT_2[1 + (q-1)T_1] & q \in C_k^u, q \in C_{q-k}^v \\ L - u & \text{otherwise} \end{cases} \tag{19}$$

where $T_1 = (u-1)/q, T_2 = (v-1)/q$.

$$m(x) = \begin{cases} \dfrac{(x^L - 1)}{(x^u - 1)\omega_{j_0}(x)\omega_{j_1}(x)\cdots\omega_{j_{q-1}}(x)} & q \in C_k^u, q \in C_{q-k}^v \\ \dfrac{(x^L - 1)}{(x^u - 1)} & \text{otherwise} \end{cases} \tag{20}$$

where $\omega_{j_i}(x) = \displaystyle\prod_{\substack{p \bmod u \in C_i^u \\ p \bmod v \in C_{j_i}^v}} (x - \alpha^p)$.

5.   Proof of Green's conjecture

Making use of the results in section 4, we can prove the conjectures on the linear complexity and feedback polynomials brought forward by D.H. Green and J. Choi.

Note that

$$S(x) = \left(\sum_{i=0}^{v-1} x^{ui}\right)\left(\sum_{i \in C_1^u} x^i\right) + \left(\sum_{i=0}^{u-1} x^{vi}\right)\left(\sum_{i \in C_1^v} x^i\right)$$
$$+ \sum_{vi \in C_1^u} x^{vi} + \sum_{ui \in C_0^v} x^{ui} \tag{21}$$

So, for $x = 1$, $S(1) \equiv 0 \bmod 2$. For $x = \alpha^j$, $j \in \Delta_1$, $S(\alpha^j) \equiv (u + 1)/2 \bmod 2$. For $x = \alpha^j, j \in \Delta_2, S(\alpha^j) \equiv (v-1)/2 \bmod 2$. For $x = \alpha^j$, $j \in \Delta_3$, $S(\alpha^j) = \sum_{i \in C_1^u}(\alpha^j)^i + \sum_{i \in C_0^v}(\alpha^j)^i$.

Fact 2 *When* $j \in \Delta_3$, $S(\alpha^j) \in \{0, 1\}$ *if and only if* $u \equiv \pm 1 \bmod 8$, $v \equiv \pm 1 \bmod 8$ *or* $u \equiv \pm 3 \bmod 8$, $v \equiv \pm 3 \bmod 8$.

When u and v such that $u \equiv \pm 1 \bmod 8$, $v \equiv \pm 1 \bmod 8$ or $u \equiv \pm 3 \bmod 8$, $v \equiv \pm 3 \bmod 8$, it follows from Fact 2 that $S(\alpha^j) \in \{0, 1\}(j \in \Delta_3)$ and

$S(\alpha^t) = 0 (t \in \Delta_3)$ for all $t \bmod u \in C_0^u$, $t \bmod v \in C_0^v$ or $t \bmod u \in C_1^u$, $t \bmod v \in C_1^v$ or $S(\alpha^t) = 0 (t \in \Delta_3)$ for all $t \bmod p \in C_0^u$, $t \bmod q \in C_1^v$ or $t \bmod p \in C_1^u$, $t \bmod v \in C_0^v$. The total numbers of $t \in \Delta_3$ are $L - u - v + 1$, the numbers of t such that $t \bmod u \in C_0^u$, $t \bmod v \in C_0^v$ or $t \bmod u \in C_0^u$, $t \bmod v \in C_1^v$ or $t \bmod u \in C_1^u$, $t \bmod v \in C_0^v$ or $t \bmod u \in C_1^u$, $t \bmod v \in C_1^v$ is $(L - u - v + 1)/4$, respectively. Hence if $u \equiv \pm 1 \bmod 8$, $v \equiv \pm 1 \bmod 8$ or $u \equiv \pm 3 \bmod 8$, $v \equiv \pm 3 \bmod 8$ then

$$\left| \{ j : S(\alpha^j) = 0, j \in \Delta_3 \} \right| = \frac{L - u - v + 1}{2}. \tag{22}$$

When u and v such that $u \equiv \pm 1 \bmod 8$, $v \equiv \pm 3 \bmod 8$ or $u \equiv \pm 3 \bmod 8$, $v \equiv \pm 1 \bmod 8$, it follows from Fact 2 that $S(\alpha^j) \notin \{0, 1\} (j \in \Delta_3)$ and $S^L(\alpha^t) \neq 0$ for all $t \in \Delta_3$. Thus in this case

$$\left| \{ j : S(\alpha^j) = 0, j \in \Delta_3 \} \right| = 0. \tag{23}$$

Then the linear complexity of all MJ-sequences can be deduced easily. For example, consider the case $u \equiv 3 \bmod 8$, $v \equiv 5 \bmod 8$, $c_L = L - 1 - (v - 1) - (u - 1) - (L - u - v + 1)/2 = (u - 1)(v + 1)/2$.

References

[1] D. Everett. Periodic digit sequences with pseudoradom properties. *GEC J.*, 33, 1966.

[2] P. Fan and M. Darnell. *Sequence Design for Communications Applications*. John Wiley, Research Studies Press, Taunton, 1996.

[3] I. Damgaard. On the randomness of Legendre and Jacobi sequences. *Advances in Cryptology: Crypto '88*, Berlin, Germany: Springer-Verlage, LNCS 403, 1990.

[4] J.H. Kim and H.Y. Song. Trace representation of Legendre sequences. *Designs, Codes Cryptogr.*, 24(3), 2001.

[5] C. Ding, T. Helleseth, and W. Shan. On the linear complexity of Legendre sequences. *IEEE Trans. Inf. Theory*, 44(3), 1998.

[6] D.H. Green and P.R. Green. Modified Jocobi sequences. *IEE Proc. Comput. Digit. Tech.*, 147(4), 2000.

[7] D.H. Green and J. Choi. Linear complexity of modified Jacobi sequences. In *IEE Proc. Comput. Digit. Tech.*, 149(3), 2002.

[8] D.H. Green, M.D. Smith, and N. Martzoukos. Linear complexity of polyphase power residue sequence. *IEE Proc. Commun.*, 149(4), 2002.

[9] D.H. Green and P.R. Green. Polyphase related-prime sequences. *IEE Proc. Comput. Digit. Tech.*, 148(2), 2001. MIT Press, Cambridge, MA, 1994.

[10] D.H. Green. Structural properties of pseudorandom arrays and volumes and their related sequences. *IEE Proc. Comput. Digit. Tech.*, 132(3), 1985.

[11] R. Lidl and H. Niederreiter. *Finite Fields.* in Encyclopedia of Mathematics and Its Applications, vol.20, Reading, MA: Assison-Wesley, 1983.

RESILIENT FUNCTIONS
WITH GOOD CRYPTOGRAPHIC PROPERTIES*

WEN Qiao-yan
1. Beijing University of Posts and Telecommunications,
Beijing, 100876, P. R. China;
2. State Key Laboratory of Information Security,
Chinese Academy of Sciences, Beijing, 100039, P. R. China
wqy@bupt.edu.cn

ZHANG Jie
Beijing University of Posts and Telecommunications,
Beijing, 100876, P. R. China;
zhj503@eyou.com

Abstract   A method of directly constructing resilient functions is presented. The functions are generated from concatenation linear functions. It is convenient to calculate the nonlinearity of the functions obtained and to discuss the algebraic degrees and propagation characteristics of them.

Keywords: correlation immune, nonlinearity, resilient function.

1.   Introductions

An (n, m, t) resilient function is an n-input m-output functions with property that it runs through every possible output -tuple an equal number of times when arbitrary inputs are fixed and the remaining inputs runs through all the input tuples once. The concept was introduced by Chor et al [1] and independently, by Bennett et al in [2]. Areas where resilient functions find their applications include fault-tolerant distributed computing, quantum cryptographic key distribution and random sequence generation for stream ciphers.

Similar to Boolean function, multi-output functions with good cryptographic properties should have the following criteria: (1)orthogonal (i.e. balance cor-

* supported by the National Natural Science Foundation of China(60373059) and the research foundation of the State key laboratory of information security

responding to Boolean function) (2)high order correlation immune (3) high
nonlinearity (4) high algebraic degree (5) propagation characteristics. Orthogonal and correlation immune are usually refered to as resiliency. These criteria
are partially opponent. It is important to discuss and harmonize them.

Up to now, there are many results about resilient functions, but for most of
them, it is difficult to discuss all these properties. In fact, most of them only
consider two properties. For example, tradeoff between correlation immunity
and nonlinearity was given by Y. Zheng and X. M. Zhang in [3], between
correlation immunity and the algebraic degree was given by Siegenthaler in
[4]. In this paper, we give a kind of construction of cryptographic resilient
functions. All these criteria above are considered. And it is easy to calculate
the nonlinearity of the functions obtained and to discuss the algebraic degrees
and propagation characteristics of them. Tradeoff among these criteria is given.
The functions are generated from concatenation linear functions. So it is more
convenient for use in practice.

2.   Preliminaries

The vector space of n tuples of elements from $GF(2)$ is denoted by V_n. These
vectors, in ascending alphabetical order, are denoted by $\alpha_0, \alpha_1, ..., \alpha_{2^n-1}$. As
vectors in V_n and integers in $[0, 2^n - 1]$ have a natural one-to-one correspon-
dence, it allows us to switch from a vector in to its corresponding integer in and
vice versa.

Let f be a function from V_n to $GF(2)$(simply, a function on V_n), The
truth table of f is a $(0,1)$-sequence defined by $(f(\alpha_0), f(\alpha_1), ...f(\alpha_{2^n-1}))$ and
the *sequence* of f is a $(1,-1)$ sequence defined by $((-1)^{f(\alpha_0)}, (-1)^{f(\alpha_0)}, ...,$
$(-1)^{f(\alpha_{2^n-1})})$. f is said to be *balanced* if its truth table assumes an equal
number of zeros and ones. We call $h(x) = c_1x_1 \oplus c_2x_2 \oplus ... \oplus c_nx_n \oplus c$ an
affine function, where $c_i(1 \leq i \leq n), c \in GF(2)$. In particular, h is named a
linear function if $c = 0$. Denote all n-variable affine function by A_n.

Functions on V_n can be considered to be a multivariate polynomial of n
coordinates. We are particularly interested in the so-called *algebraic normal
form* representation in which a function is viewed as the sum of products of
coordinates. The *algebraic degree* $deg(f)$ of a function f is the number of
coordinates in the longest product in the algebraic normal form. The *hamming
weight* of a vector v is the number of ones in v . Let f and g be two functions on
V_n , the *hamming distance* of them is the number of distinct elements between
their sequence, denoted by $d(f,g)$. The *nonliearity* of f is defined by $N_f = \min_{g \in A_n} d(f,g)$

If denote the sequences of f and g by ξ_f and ξ_g , respectively, then $d(f,g) = 2^{n-1} - \frac{1}{2} < \xi_f, \xi_g >$ [5, lemma 6]. So we have $N_f = 2^{n-1} - \frac{1}{2} \max_{l \in A_n} < \xi_f, \xi_g >$

It is well known that the nonlinearity of f on V_n satisfies $N_f \leq 2^{n-1} - 2^{\frac{1}{2}n-1}$.

f is said to satisfies the *propagation criterion* with respect to a non-zero vector α in V_n , if $f(x) \oplus f(x + \alpha)$ is a balanced function. Furthermore, it satisfies the propagation criterion of degree k if it satisfies the propagation criterion with respect to all $\alpha \in V_n$ with $1 \leq W(\alpha) \leq k$.

A Boolean function $f(x)$ on n variables is said to be $m - th$ order correlation immune $(m - CI)$, if for any m-tuple of independent identically distributed binary random variables $X_{i_1}, X_{i_2}, ..., X_{i_m}$. we have $I(X_{i_1}, X_{i_2}, ..., X_{i_m}; Z) = 0, 1 \leq i_1 < i_2 < ... < i_m \leq n$. where $Z = f(X_1, X_2, ..., X_n)$ and $I(X; Z)$ denotes the mutual information[7].

Corresponding to Boolean function, we define concepts of multi-output function. Let $F = (f_1, f_2, ..., f_m)$ is a function from V_n to V_m, its nonlinearity is defined as the minimum among the nonlinearities of all nonzero linear combinations of component functions of F. i.e.

$$N_f = \min_g \{N_g | g = \bigoplus c_i f_i, c_i \in GF(2), (c_1, c_2, ..., c_m) \neq (0, 0, ..., 0)\}$$

The algebraic degree of F , denoted by $deg(F)$, is defined as the minimum among the algebraic degrees of all nonzero linear combinations of the component functions of F , namely,

$$deg(f) = \min_g \{deg(g) | g = \bigoplus c_i f_i, c_i \in GF(2),$$
$$(c_1, c_2, ..., c_m) \neq (0, 0, ..., 0)\}$$

F is called to satisfy the propagation criterion of degree k if its all nonzero linear combination satisfies the propagation criterion with respect to all $\alpha \in V_n$ with $1 \leq W(\alpha) \leq k$.

Definition 1: Let $F = (f_1, f_2, ..., f_m)$ be a function from V_n to V_m, where $n \geq m \geq 1$, and let $x = (x_1, x_2, ..., x_n) \in V_n$.

1)F is said to be *unbiased* with respect to a fixed subset $T = \{j_1, ..., j_t\}$ of $\{1, ..., n\}$, if for every $(\alpha_1, \alpha_2, ..., \alpha_t) \in V_t$, $(f_1(x), ..., f_m(x))|_{x_{i_1}=\alpha_1,...,x_{i_t}=\alpha_t}$ runs through all the vectors in V_m each 2^{n-m-t} times while $(x_{i_1}, ..., x_{i_{n-t}})$ runs through V_{n-t} once, where $t \geq 0$, $\{i_1, ..., i_n - t\} = \{1, ..., n\} - \{j_1, ..., j_t\}$ and $i_1 < i_2 < ... < i_{n-t}$.

2) F is said to be a t-resilient function if F is unbiased with respect to every subset T of $\{1, ..., n\}$ with $|T| = t$.

The parameter t is called the *resiliency* of the function.

3.   Previous constructions and results

Given any vector $\delta = (i_1, i_2, ..., i_s) \in V_s$, we define a function on V_s by

$$D_\delta(y) = (y_1 \oplus \bar{i}_1)(y_2 \oplus \bar{i}_2)...(y_s \oplus \bar{i}_s)$$

where $y = (y_1, y_2, ..., y_s)$ and $\bar{i} = 1 \oplus i$ indicates the complement of i. The addition and multiplication are over $GF(2)$. Obviously, $D_\delta = 1$ if and only if $y = \delta$.

Suppose that $\Phi_{n,s} = \{ \Psi_{0...0}, \Psi_{0...1}, ..., \Psi_{1...1}, \}$ is a set containing 2^{n-s} linear functions on V_s, each is indexed by a vector in V_{n-s}. $\Phi_{n,s}$ can be a multi-set.

Theorem 1[5] Let n and s be positive integers with $n > s$, $x = (x_1, x_2, ..., x_s)$, $y = (y_1, y_2, ..., y_{n-s})$ and r is an arbitrary function on V_{n-s}. Set

$$g(y, x) = \bigoplus_{\delta \in V_{n-s}} D_\delta(y) \Psi_\delta(x) + r(y)$$

then $g(y, x)$ is a balanced kth-order correlation immune function on V_n, where k is an integer satisfying $k \geq \min\{W(\gamma_\delta)|\delta \in V_{n-s}\} - 1$, $\Psi_\delta(x) = <\gamma_\delta, x> \in \Phi_{n,s}$ and $\gamma_\delta \in V_s$.

Next, we discuss cryptographic criteria of function g given above.

Theorem 2[5] Let n and s be positive integers with $n > s > 2$, t_δ is the number of times that a linear function $\Psi_{delta}(x)$ appears in $\Phi_{n,s}$. Let $t = max\{t_\delta|\delta \in V_{n-s}\}$, then $N_g \geq 2^{n-1} - t2^{s-1}$.

Theorem 3[5] Let k, n and s be integers with $k \geq 1$ and $n > s \geq k + 2$, then a balanced kth-order correlation immune function on V_n of algebraic degree $n - s + 1$ can be obtained.

Theorem 4[5] If all φ_δ are distinct linear functions on V_s, then g satisfies the propagation criterion with respect to all $\gamma = (\beta, \alpha)$, $0 \neq \beta \in V_{n-s}$ and $\alpha \in V_s$.

Theorem 5[6] Let $F = (f_1, f_2, ..., f_m)$ be a function from V_n to V_m, where n and m are integers with $n \geq m \geq 1$ and each f_j is a function on V_n. Then F is a (n, m, t)-resilient function if and only if every nonzero linear combination of $f_1, f_2, ..., f_m$, $f(x) = \bigoplus_{i=1}^m c_i f_i(x)$, is a $(n, 1, t)$-resilient function, where $x = (x_1, x_2, ..., x_n) \in V_n$.

4.   New construction of resilient functions

For integers k and n with $0 \geq k < n$, Let $\Omega_{k,n}$ denote the set of linear functions on V_n that have $k + 1$ or more non-zero coefficients, namely $\Omega_{k,n} = \{\varphi|\varphi(x) = <\beta, x>, x, \beta \in V_n, W(\beta) \geq k + 1)\}$.

Let n, m, t be integers with $n \geq m$, select $m2^{n-s}$ (repetition is permitted in the selection) functions from $\Omega_{t,s}$ and separate them into m groups arbitrarily, denoted m groups by $\Psi_1, \Psi_2, ..., \Psi_m$. Each group has 2^{n-s} functions. Denote $\Psi_i = \{\varphi_{i,\delta_j}|\delta_j \in V_{n-s}\}, 1 \geq i \geq m$, Select m functions on V_{n-s} arbitrarily, denoted by $r_1(y), r_2(y), ..., r_m(y)$, separately.

Let $g_i(y, x) = \bigoplus_{\delta_j \in V_{n-s}} D_{\delta_j} \varphi_{i,\delta_j}(x) \oplus r_i(y)$ where $\varphi_{i,\delta_j} \in \Psi_i$ and $1 \geq i \geq m$. Set

$$F(y, x) = (g_1(y, x), g_2(y, x), ..., g_m(y, x) \ (*)$$

where $y \in V_{n-s}$ and $x \in V_s$.

Theorem 6 $F(y, x)$ constructed above is an (n, m, t) -resilient function.

Proof. Consider an arbitrary nonzero linear combination of the component functions of $P(z)$, say $P(z) = \bigoplus_{i=1}^{m} c_i g_i(y, x), c_i \in GF(2)$. By Theorem 1, each g_i is an balanced tth-order correlation immune function, i.e. is a $(n, 1, t)$-resilient function. By Theorem 5, F is an (n, m, t)-resilient function.

Theorem 7 Let n, s be integers with $n > s > 2$, Denote by q_i the maximal number of times a linear functions $\varphi_{i,\delta_i}(\delta_i \in V_{n-s})$ appears in Ψ_i, $1 \leq i \leq m$, and let $q = max\{q_i | 1 \leq i \leq m\}$. Then the nonlieariety of function obtained from (*) is $N_f \geq 2^{n-1} - p2^{s-1}$.

Proof. Denote the sequences of $g_i(y, x)$ by ξ_{g_i}, $1 \leq i \leq m$. Let $P(z) = \bigoplus_{i=1}^{m} c_i g_i(y, x)$, then the sequence of $P(z)$ $\xi = c_1 \xi_{g_1} \oplus ... \oplus c_m \xi_{g_m}$. Select an affine functions $l(x)$ on V_n arbitrarily, ξ_l denote its sequence. Then

$$< \xi, \xi_l > = < c_1 \xi_{g_1} \oplus ... \oplus c_m \xi_{g_m}, \xi_l > \leq max\{< \xi_{g_i}, \xi_l > | 1 \leq i \leq m\}.$$

by theorem 2, therefore $N_F \geq max_{1 \leq i \leq m} = 2^{n-1} - p2^{s-1}$. So $N_F \geq 2^{n-1} - p2^{s-1}$ according to the definition of nonlinearity.

Corollary 1: Let n, m, t be integers with $n > m$, if there exits an integer s, such that

$$\binom{s}{t+1} + \binom{s}{t+2} + ... + \binom{s}{s} > m2^{n-s}$$

then the nonlieariety of function obtained from (*) is $N_F \geq 2^{n-1} - 2^{s-1}$.

Next, we discuss the algebraic of function obtained by our method.

$$deg(F) = \min_{h}\{deg(h) | h = \bigoplus_{i=1}^{m} c_i g_i, c_i \in GF(2), (c_1, ..., c_m) \neq (0, ..., 0)\}$$

Theorem 8 If $|\Omega_{t,s}| \geq m2^{n-s}$, then the algebraic of function obtained from (*) is $n - s + 1$. When $s = t + 1$, the function achieve the maximum algebraic degree $m - t - 1$.

Proof. Arrange the functions in $\Omega_{t,s}$ on the length and footnote of variable ascending alphabetical order. Select $m2^{n-s}$ functions from $\Omega_{t,s}$ in order from beginning and separate then into m sets. Denote the sets $\Psi_1, \Psi_2, ..., \Psi_m$. Then construct multi-output function by method (*). It is easy to prove the algebraic degree of the arbitrary nonzero linear combination of g ($1 \leq i \leq m$) is $n - s + 1$. By the definition of F, we have $deg(F) = n - s + 1$.

Theorem 9 In the construction (*), if each $\Psi_i(1 \geq i \geq m)$ are not multiset, then $F(y, x)$ satisfies the propagation criterion with respect to all γ with $\gamma = (\alpha, \beta)$, $\alpha \in V_{n-s}, \beta \in V_s$ and $\alpha \neq 0$.

Proof. For any arbitrary $\gamma = (\alpha, \beta)$, $\alpha \in V_{n-s}$, $\beta \in V_s$ and $\alpha \neq 0$ and nonzero linear combination $(c_1, c_2, ..., c_m)$ of F, let $z = (y, x)$, then we have:

$$\sum_{i=1}^{m} c_i \bigoplus \sum_{i=1}^{m} c_i g_i(z\gamma) = \sum_{i=1}^{m} c_i(g_i(z) \oplus g_i(z + \gamma)),$$

by [5,theorem 5], $g_i(z) \oplus g_i(z \oplus \gamma)$ is balance. Therefore $\sum_{i=1}^{m} c_i(g_i(z) \oplus g_i(z \oplus \gamma))$ is balance too. i.e. F satisfies the propagation criterion with respect to γ.

5.   Example

We construct $n = 11, m = 3, t = 3$, i.e. $(11, 3, 3)$-resilient function. Select $s = 7$.

$$\binom{7}{4} + \binom{7}{5} + \binom{7}{6} + \binom{7}{7} = 64 > 3 \times 16 = m2^{n-s}$$

For convenient, we denote linear function $x_{i_1} \oplus x_{i_2} \oplus ... \oplus x_{i_k}$ as number sequence $i_1 i_2 ... i_k$. Let

$$\Psi_1 = \begin{pmatrix} 1234 & 1235 & 1236 & 1237 & 1245 & 1246 & 1247 & 1256 \\ 1257 & 1267 & 1345 & 1346 & 1347 & 1356 & 1357 & 1367 \end{pmatrix}$$

$$\Psi_2 = \begin{pmatrix} 1456 & 1457 & 1467 & 1567 & 2345 & 2346 & 2347 & 2356 \\ 2357 & 2367 & 2456 & 2457 & 2467 & 2567 & 3456 & 3457 \end{pmatrix}$$

$$\Psi_3 = \begin{pmatrix} 3467 & 3567 & 4567 & 12345 & 12346 & 12347 & 12356 & 12357 \\ 12367 & 12456 & 12457 & 12467 & 12567 & 13456 & 13457 & 13467 \end{pmatrix}$$

Select three functions $p_i(y, x)(i = 1, 2, 3)$ from V_{11} arbitrarily. By the method of above, we get $g_i(y, x)(i = 1, 2, 3)$ as following:

$$g_1(y, x) = y_1 y_2 y_3 y_4 (x_4 \oplus x_5 \oplus x_6 \oplus x_7) \oplus p_1(y, x)$$

$$g_2(y, x) = y_1 y_2 y_3 y_4 (x_4 \oplus x_5) \oplus p_2(y, x)$$

$$g_3(y, x) = y_1 y_2 y_3 y_4 (x_1 \oplus x_3 \oplus x_4 \oplus x_6 \oplus x_7) \oplus p_3(y, x)$$

Now the multi-output function $F(y, x) : V_{11} \to V_3$ is obtained as: $F(y, x) = (g_1, g_2, g_3)$, Obviously, F is an resilient function and resiliency $t = 3$, $deg(F) = 5$, $N_f = 2^{10} - 2^6$. F satisfies the propagation criterion with respect to all $\gamma = (\beta, \alpha), 0 \neq \beta \in V_4, \alpha \in V_7$.

6. Conclusion

We have studied the resilient functions using concatenation of the linear functions. The resilient functions obtained by our method have good cryptographic properties. In particular, it is convenient to calculate the nonlinearity of functions obtained and discuss their algebraic degrees and propagation characteristics. This direct construction from concatenation linear functions is more convenient for use in practice.

References

[1] B.Chor, O. Goldreich, J. Hastad, J.Friedman, S. Rudich and R. Smolensky, The bit extraction problem or t-resilient functions. IEEE symposium on Foundations of Computer Science 26(1985), 396-407.

[2] C. H. Bennett, G. Brassard and J. M. Robert, Privacy amplification by public discussion. SIAM J.computing 17(1988),210-229.

[3] Y. L. Zheng and X. M. Zhang, Improved upper bound on the nonlinearity of high order correlation immune functions. In Selected Areas in Cryptography-SAC 2000(Lecture Notes in Computer Science). Berlin, Germany: Springer-Verlag, 2000, Vol. 2012, pp. 49-63.

[4] T. Siegenthaler, Decrypting a class of stream cipher using ciphertext only. IEEE Transactions on Computers 34(1) (1985)81-85.

[5] J.Seberry, X. M. Zhang and Y. L. Zheng, On constructions and nonlinearity of correlation immune functions.

[6] X. M. Zhang and Y. L. Zheng, Cryptographically resilient functions, IEEE Transactions on Information Theory, Vol. 43, No. 5 Sept.1997, pp.1740-1747.

[7] T. M. Cover, J. A. Thomas, "Elements of information theory", John Wiley & Sons Inc., 1991.

DIFFERENTIAL FACTORING FOR INTEGERS*

Chuan-Kun Wu

Key Lab of Inform. Security, Institute of Software, Chinese Academy of Sciences, Beijing, China
kwu@is.iscas.ac.cn

Abstract   This paper presents two new factoring methods which apply to numbers with certain properties. When one factor of an integer has a long all-zero or all-one string in its binary representation, factorization of the integer can be made more efficient using one of the complementary algorithms proposed in this paper. Based on the proposed algorithms, new criterion for secure RSA moduli should be taken into account.

Keywords: differential factorization algorithm, new criterion for secure RSA moduli

1.   Introduction

In 1978, R.L.Rivest, A.Shamir and L.Adleman [5] proposed a public key cryptosystem based on the intractability of factorization. After 20 years worldwide study and analysis of the cryptosystem, it is believed that breaking RSA is as hard as integer factorization, although the equivalence of factorization and the security of RSA has not been strictly proved yet. There have been many different good factorization method (see for example [2, 3]). Each method works for integers with certain properties more efficiently than others. For example, for an RSA modulus $n = pq$, the $(p-1)$-method can work well mostly to find a prime factor p of n if $p-1$ has only small prime factors. Against this method, the concept of "safe" modulus of RSA was proposed, which says: the modulus n should be the product of such two primes p and q that both $p-1$ and $q-1$ must have a large prime factor. An integer n is said to have a *large prime factor* p if p is a factor of n and $p >> \sqrt{n}$. If $(p-1)/2$ is a prime number as well, p is called a *perfect prime*. In this paper we propose two new factoring methods which are complementary to each other. These methods work efficiently for numbers when one of their factors is very close to $t \cdot 2^t$ for some t and n while the other factors are not too large. We further show that some perfect primes are not even safe when they are used in RSA.

* Supported by Natural Science Fundation of China No. 90304007.

2.   Right shifting and its properties

Let n be a non-negative integer. Define by $R(n) = \lfloor \frac{n}{2} \rfloor$, the integral part of $\frac{n}{2}$. It is known that $R(n)$ is equivalent to right shifting of integer n in its binary representation. We denote by $R^k(n)$ the k-time repeat of right shifting function on n. Denote by $l(n) = \lfloor \log_2 n \rfloor + 1$, the length of n in its binary representation. Then the following properties can be verified easily:

P1: $R(n) = 0$ for $n \leq 1$.
P2: $R^{l(n)}(n) = 0$, $R^{l(n)-1}(n) = 1$.
P3:   If $2|n$, i.e., n is even, then for any integer a, $R(an) = aR(n)$.
P4: $R(ab) \geq 2R(a)R(b)$. Equality holds if and only if $a = b = 1$ or $2|a, 2|b$.
P5: $R(ab) \geq aR(b)$. Equality holds if and only if $2|n$ or $a \in \{0, 1\}$.
P6: $R(a + b) = R(a) + R(b)$ if $2|ab$, and $R(a + b) = R(a) + R(b) + 1$ otherwise.
P7: $R(a - b) = R(a) - R(b) + 1$ if a is even and b is odd, and
          $R(a - b) = R(a) - R(b)$ otherwise.
P8: $R^k(n) = \lfloor \frac{n}{2^k} \rfloor$.

For a positive integer n and an odd integer d, define another function

$$\Gamma_d(n) = \begin{cases} \frac{n}{2} & \text{if } 2|n, \\ \frac{n-d}{2} & \text{else.} \end{cases}$$

It is seen that $\Gamma_1(n) = R(n)$. So the function $\Gamma_d(n)$ is a generalization of $R(n)$. The relationship between $\Gamma_d(n)$ and $R(n)$ can be expressed as

$$\Gamma_d(n) = R(n) + \frac{(-1)^n - 1}{2} R(d) = R(n) - n_0 R(d),$$

where n_0 is the least significant bit of n in binary representation. Now we determine how fast the repeat applying Γ_d on n decreases it to 0 or less. Denote by $n^{(0)} = n$, $n^{(i)} = \Gamma_d(n^{(i-1)})$, $i = 1, 2, \cdots$. For a randomly given n, each $n^{(i)}$ in the sequence $\{n^{(i)}\}$ is equally likely to be 0 or 1. So the expected value of $\Gamma_d(n)$ is $\overline{\Gamma_d}(n) = (\frac{n}{2} + \frac{n-d}{2})/2 = \frac{n}{2} - \frac{d}{4}$. This gives a recurrence formula $n^{(i)} = \frac{n^{(i-1)}}{2} - \frac{d}{4}$. By this recursion we can deduce

$$n^{(k)} = \frac{n}{2^k} - \frac{d}{2^2} - \frac{d}{2^3} - \cdots - \frac{d}{2^{k+1}} \approx \frac{n}{2^k} - \frac{d}{2}.$$

Set $n^{(k)} = 0$ we have $k \approx \log_2 \frac{2n}{d}$. Take the integral value $k = \lceil \log_2 \frac{n}{d} \rceil + 1$, we have

Lemma 2.1. *Given an integer n and an odd integer d at random. Then repeat applying function Γ_d on n for $\gamma_d(n) = \lceil \log_2 \frac{n}{d} \rceil + 1$ times will expect to yield a value ≤ 0, i.e., $\Gamma_d^{\gamma_d(n)}(n) \leq 0$.*

3.   An algorithm

We introduce the following algorithm and analyse its performance.

Algorithm $\mathcal{A}(n, d)$

> **Input:** Odd numbers n and d.
> 1. $a = n$.
> 2. $a = \Gamma_d(a)$.
> 3. **If** $2|a$, set $g = \gcd(a - d, n)$; **else** set $g = \gcd(a, n)$.
> 4. **If** $g|n$, output g and stop.
> 5. **If** $a > 1$ goto step 2; **else** report failure and exit.

Theorem 3.1. *Let* $n = pq$, *where* p *and* q *do not have to be primes. If there exist integers* k *and* d *such that* $2^k|(q - d)$ *and* $p < 2^{k+1}$, *then the algorithm* $\mathcal{A}(n, d)$ *will yield a nontrivial factor of* n.

Proof: By writing $n = (t \cdot 2^k + d)p$ and the properties of $\Gamma_d()$ and $R()$ we know that for $i < l(p) \le k$, we have

$$\Gamma_d^i(n) = p \cdot t \cdot R^i(2^k) + d \cdot R^i(p).$$

Let $i = l(p) - 1$, then by property **P2** we have $a = \Gamma_d^i(n) = p \cdot t \cdot 2^{k-i} + d$. If $t \cdot 2^{k-i}$ is odd, i.e., t is odd and $i = k$, then a is even, and in the 3-rd step of the algorithm we get $g = \gcd(tp, n)$. Since $tp < n$ and $p|g$, we get a nontrivial factor g of n. If $t \cdot 2^{k-i}$ is even, then a is odd. Applying $\Gamma_d(a)$ one more time we get a new value $a = p \cdot t \cdot 2^{k-i-1}$. At this stage, regardless whether $t \cdot 2^{k-i-1}$ is odd or even, the algorithm will finally find $a = pt'$, where t' is odd. So in step 3 of the algorithm we can certainly find a nontrivial factor $g = \gcd(pt, n)$ of n. □

If there exits a d so that algorithm $\mathcal{A}(n, d)$ can find a proper factor of n, then d is called a *differentia of* n *associated with algorithm* $\mathcal{A}(n, d)$, or a-differentia. Apparently the one satisfying the properties of theorem 3.1 is an a-differentia of n, but not necessarily the smallest one.

It is easy to verify that within at most $l(q) - 1$ rounds, the algorithm will terminate. The problem now is how to find a differentia without the knowledge of p and q. One way to achieve this is to set $d = 1$ and execute algorithm $\mathcal{A}(n, 1)$. If it fails to find a factor of n, increase d by 2 and execute $\mathcal{A}(n, d)$ again. This is not efficient for factoring general integers where even the smallest differentia is very large. Note that even if the theoretical d satisfying the properties of theorem 3.1 has to be very large, it is possible that there exists another a-differentia d of n much smaller than the theoretical value.

Let p and q be in binary representationand assume that there exist k and $d < q$ such that $2^k|(q - d)$ and $p < 2^{k+1}$. Then the length of p in its binary representationis no longer than that of q, i.e., $l(p) \le l(q)$. It is noticed that the $l(p) - 1$ least significant bits of q is the theoretical d which guarantees the factorization using the above algorithm. In order for the algorithm to be efficient, the $l(p) - 1$ least significant bits of q should be a long string of zeros before a nonzero bit appears so that d can be reasonably small.

4.   A complementary algorithm

Define
$$\Delta_d(n) = \begin{cases} \frac{n}{2} & \text{if } 2|n, \\ \frac{n+d}{2} & \text{else.} \end{cases}$$

Denote by $n^{(0)} = n$, $n^{(i)} = \Delta_d(n^{(i-1)})$. Then it can be assumed in general that $n^{(i)}$ is equally likely to be 0 or 1. So given a random $n^{(i-1)}$, the expected value of $n^{(i)}$ is
$$n^{(i)} = \Delta_d(n^{(i-1)}) = \frac{n^{(i-1)}}{2} + \frac{d}{4}.$$

With this recursion we have
$$n^{(i)} = \frac{n}{2^k} + \frac{d}{2^2} + \frac{d}{2^3} + \cdots + \frac{d}{2^{k+1}} \approx \frac{n}{2^k} + \frac{d}{2}.$$

Set $n^{(i)} = d$, we have $k \approx \log_2 \frac{2n}{d}$. So we have

Lemma 4.1. *Given an integer n and an odd integer d at random. Then repeat applying function Δ_d on n for $\gamma_d(n) = \lceil \log_2 \frac{n}{d} \rceil + 1$ times will expect to yield a value $\leq d$, i.e., $\Delta_d^{\gamma_d(n)}(n) \leq d$.*

Note that when $\Delta_d^k(n)$ gets a value less than or equal to d, further applying Δ_d on it may not decrease its value at all. This is one of the differences between Δ_d and Γ_d which should be taken into account in algorithm design.

Similar to algorithm $\mathcal{A}(n, d)$ we can develop a complementary algorithm as follows:

Algorithm $\mathcal{B}(n, d)$
> **Input:** Odd numbers n and d.
> 1. $a = \Delta_d(n)$.
> 2.   If $2|a$, $g = \gcd(a + d, n)$; **else** $g = \gcd(a, n)$.
> 3.   If $g|n$, output g and stop.
> 4.   If $a > d$, let $a = \Delta_d(a)$ and goto step 2;
> **else** report failure and stop.

Theorem 4.2. *Let $n = pq$, where p and q do not have to be primes. If there exist integers k and d such that $2^k|(q + d)$ and $p < 2^{k+1}$, then the algorithm $\mathcal{B}(n, d)$ will yield a nontrivial factor of n.*

Proof. The proof of the theorem is similar to that of theorem 3.1 by noticing $\Delta_d^i(n) = p \cdot t \cdot R^i(2^k) - d \cdot R^i(p)$.   □

If there exits a d such that algorithm $\mathcal{B}(n, d)$ can find a proper factor of n, then d is called a *differentia of n associated with algorithm $\mathcal{B}(n, d)$*, or b-differentia. Those values of d satisfying the properties of theorem 4.2 are b-differentiae. Similar to the case of a-differentiae, there might exit a b-differentia much smaller than those satisfying the properties of theorem 4.2.

Note: In the factorization of $n = pq$ using algorithm $\mathcal{A}(n, d)$, d has to be smaller than q. However, if algorithm $\mathcal{B}(n, d)$ is used, d could be anything provided that it satisfies the properties of theorem 4.2. But in practical implementation, d is normally initialized by a small odd integer, then increased by 2 in each trial. So it cannot be very large. Nevertheless, algorithm $\mathcal{B}(n, d)$ allows multiple choices of d (d must be less than n in any circumstances). Once we find a correct d, we can get a factor of n.

Note: Like algorithm $\mathcal{A}(n, d)$, we cannot figure out what kind of integers are vulnerable to algorithm $\mathcal{B}(n, d)$. But theorem 4.2 tells that if one of the factors of n has a long all-one string in its binary representation, algorithm $\mathcal{B}(n, d)$ works efficiently by initializing $d = 1$ and increased by step 2.

5.   Some *perfect* primes are not perfect

Pollard's $(p - 1)$-factoring method works efficiently when $p - 1$ has only small prime factors, where p is a prime factor of n. As an impact of this method on RSA, it was suggested to use *strong* primes in RSA. A prime p is said strong if $p - 1$ has a large prime factor. Rivest [6] further restrict the condition as: Such a prime p should be used in RSA that $p - 1$ has a large prime factor p' and $p' - 1$ also has a large prime factor. Based on the algorithm $\mathcal{A}(n, d)$, when a prime has particular properties, even if it satisfies Rivest's condition, a composite with the prime as a factor can be factorized very easily. For examples, $p = 16421$ is a prime, $p' = 2 \times p + 1 = 32843$ and $p'' = 2 \times p' + 1 = 65687$ are primes as well. For any integer $n = p''q$ with $q < 131072$, using algorithm $\mathcal{A}(n, d)$ it can be factorized by choosing $d = 151$. This is not too hard if we initialize $d = 1$ and let it increase by 2 in every round. Although p'' is by the conventional knowledge known as *perfect prime*, and satisfies Rivest condition, and it is even a prime applicable to Rabin cryptosystem [4], it is not safe if used in RSA or Rabin cryptosystem.

In the implementation of RSA, if the primes are chosen at random, then the algorithms above do not work effectively. However, there are no methods to efficiently determine whether a given large number is prime in general case. The most acceptable method is probabilistic method [7, pages 129-138]. Although it can give us as high confidence as we wish to believe whether a number is prime, there is still a possibility that some people would still use particular classes of primes where their primality can be completely determined. There is a way to determine whether a number in the form $2^q - 1$ is a prime, where q is a smaller prime [1, Vol.2, page 409]:

Lemma 5.1. *Let q be odd prime. Define sequence $< L_n >$ by:*

$$L_0 = 4, \quad L_{n+1} = (L_n^2 - 2) \bmod (2^q - 1).$$

Then $2^q - 1$ is prime if and only if $L_{q-1} = 0$.

Based on lemma 5.1, if a prime in the form $2^q - 1$ is used in RSA or similar systems where the security is based on the hardness of integer factorization, then the system is insecure as an integer having such a prime factor can be factorized using algorithm $\mathcal{B}(n, 1)$ at no cost (the first round is successful) provided that the remaining factor is not larger than 2^{q+1}. So, the algorithms in this paper further address that practical primes used for designing public key cryptosystems should be chosen at random.

6.   Preprocessing for parallel computation

Let $q = t \cdot 2^k + d$, $n = pq = pt \cdot 2^k + dp$. If d can be factorized into $d = d_1 d_2$, then we can write $n = pt \cdot 2^k + d_1 d_2 p$. With a similar analysis as the proof of theorem 3.1 we know that if $d_2 p < 2^{k+1}$, then using algorithm $\mathcal{A}(n, d_1)$ will be able to find a proper factor of n and hence d_1 is an a-differentia. We may find other a-differentiae using different factorization of d when p is sufficiently small. However when p is of similar size as q, or d is a prime as well, this method does not work. So we need some other techniques.

Let $n_1 = \Gamma_{d_0}(n) = pt \cdot 2^{k-1} + \frac{dp - d_0}{2}$. Denote by $p_0 = \frac{dp - d_0}{2}$. If p_0 can be factorized into $p_1 = d_1 p_1$, where $p_1 < 2^k$ and d_1 is odd, then further applying algorithm $\mathcal{A}(n, d_1)$ on n_1 will be successful in finding a proper factor of n. Note that in this case the routine in algorithm $\mathcal{A}(n, d_1)$ should be revised so that it starts with n_1 instead of n itself. This preprocessing for n gives the following advantages:

- Preprocessing for n may yield a smaller d_1 which enables algorithm $\mathcal{A}(n, d_1)$ to find a proper factor of n while the smallest a-differentia of n is larger than d_1.

- As different value of d_0 in the preprocessing may result in totally different outcomes, parallel computation is made possible by taking different initial values of d_0.

- When the value of d_0 is sufficiently large, further using algorithm $\mathcal{B}(n, d_1)$ on n_1 may be more efficient than using algorithm $\mathcal{A}(n, d_1)$.

Note: When n has small a-differentia, preprocessing may lead to a worse result. So when implementing parallel computation, at least one computation is devoted to the direct algorithm $\mathcal{A}(n, d)$.

Similar preprocessing techniques can be developed for the algorithm $\mathcal{B}(n, d)$. In contrary to that of algorithm $\mathcal{A}(n, d)$, a value d_0 should be added to n instead of subtracted from n. When the value of d_0 is sufficiently large, using algorithm $\mathcal{A}(n, d_1)$ may be more efficient than using algorithm $\mathcal{B}(n, d_1)$ in the forthcoming computation. So we can develop the following algorithm with multiple routines.

Algorithm $\mathcal{AB}(n, d_0, B)$

    Input: Odd numbers n, d_0 and B.

    Properties: n is the integer to be factorized; d_0 is an arbitrary integer for preprocessing; B is an upper bound of the algorithm. Preprocessing: $n_1 = (n - d_0)/2$; $n_2 = (n + d_0)/2$.

    while $2|n_1$ **do** $n_1 = n_1/2$. **while** $2|n_2$ **do** $n_2 = n_2/2$.

    Routine 1:

    for $d = 1$ **to** B **with step 2 do** $\mathcal{A}(n, d)$.

    Routine 2:

    for $d = 1$ **to** B **with step 2 do** $\mathcal{B}(n, d)$.

    Routine 3:

    for $d = 1$ **to** B **with step 2 do**

    3.1 $a = \Gamma_d(n_1)$.

    3.2   **If** $2|a$, $g = \gcd(a - d, n)$; **else** $g = \gcd(a, n)$.

    3.3   **If** $g|n$, output g and stop.

    3.4   **If** $a > 1$, let $a = \Gamma_d(a)$ and goto step 3.2;

            else report failure and exit.

    Routine 4:

    for $d = 1$ **to** B **with step 2 do**

    4.1 $a = \Delta_d(n_1)$.

    4.2   **If** $2|a$, $g = \gcd(a + d, n)$; **else** $g = \gcd(a, n)$.

    4.3   **If** $g|n$, output g and stop.

    4.4   **If** $a > d$, let $a = \Delta_d(a)$ and goto step 4.2;

            else report failure and exit.

    Routine 5:

    for $d = 1$ **to** B **with step 2 do**

    5.1 $a = \Gamma_d(n_2)$.

    5.2   **If** $2|a$, $g = \gcd(a - d, n)$; **else** $g = \gcd(a, n)$.

    5.3   **If** $g|n$, output g and stop.

    5.4   **If** $a > 1$, let $a = \Gamma_d(a)$ and goto step 5.2;

            else report failure and exit.

    Routine 6:

    for $d = 1$ **to** B **with step 2 do**

    6.1 $a = \Delta_d(n_2)$.

    6.2   **If** $2|a$, $g = \gcd(a + d, n)$; **else** $g = \gcd(a, n)$.

    6.3   **If** $g|n$, output g and stop.

    6.4   **If** $a > d$, let $a = \Delta_d(a)$ and goto step 6.2;

            else report failure and exit.

It is noted that algorithm $\mathcal{AB}(n, d_0, B)$ can further be implemented in parallel by feeding different values for d_0. Further digging up of the algorithm may include multiple preprocessing, i.e., after the preprocessing for n we get n_1, then further preprocess n_1 we get n_2, continue this procedure for t times to get n_t, and then use n_t in routines 3 and 4 instead of n_1.

7.   A few small examples

In this section we demonstrate a few small examples to illustrate how the algorithms work. We denote by d^- the smallest theoretical a-differentia (as in theorem 3.1) and d^+ the smallest theoretical b-differentia (as in theorem 4.2) of a given number n. (d_0^-, d_1^-) means routine 3 is successful and (d_0^+, d_1^+) means routine 6 is successful, where d_0 (d_0^- or d_0^+) is used in preprocessing, and d_1 (d_1^- or d_1^+) is the differentia when factoring from the appropriate routine gets a nontrivial factor. Routine 4 and 5 have not been tested.

n	9797	91961033	5821730755786530029567
p	97	9221	2147483647
q	101	9973	2710954639361
d^-	33	1029	830275585
d^+	27	6411	243466239
(d_0^-, d_1^-)	(1, 5); (3, 7)	(5, 45); (23, 27)	(1, 20853); (7, 16385)
(d_0^+, d_1^+)	(1, 11); (3, 17)	(1, 207); (9, 323)	(17, 917); (39, 851)

Table 1: Preprocessing simplifies the factorization.

From table 1 we can see that with the preprocessing, we may be able to find a d_0 which is much smaller than the a-differentia and the b-differentia of n, and after the preprocessing with d_0 we can find a proper factor of n much easier.

8.   Concluding remarks

In this paper we have developed two new methods for factoring integers. They are efficient for integers with particular properties. It is seen that one class of those integers in that one of their factors (not necessarily prime factors) has a long segment of all-zero or all-one string in its binary representation.

The idea for preprocessing is that, in case one of the factors of n has a larger number of zeros compared with the number of ones (or vice versa) in its binary representation, but not a segment of all-zero (or all-one) string, the preprocess would hopefully join the strings into a longer one and consequently one of the complementary algorithms works.

References

[1] D.E. Knuth, *The Art of Computer Programming*, 3rd ed., Addision-Wesley, 1997.

[2] J.M.Pollard, Theorems on factorization and primality testing, *Proc. of Cambridge Philos. Soc.*, Vol.76, 1974.

[3] C. Pomerance, The Quadratic Sieve Factoring Algorithm",*EUROCRYPT'84*, 1985.

[4] M.O. Rabin, Digitized signatures and public-key functions as intractable as factorization, *MIT Laboratory for Computer Science Technical Report*, LCS/TR-212, 1979.

[5] R.L.Rivest, A.Shamir, and L.Adleman, A method for obtaining digital signatures and public-key cryptosystems, *Comm. ACM*, Vol.21, No.2, 1978.

[6] R.L.Rivest, Remarks on a proposed cryptanalytic attack on M.I.T. public-key cryptosystem, *Cryptologia*, Vol.2, 1978.

[7] D.R. Stinson, *Cryptography, Theory and Practice*, CRC Press, 1995.

SIMPLE AND EFFICIENT SYSTEMATIC A-CODES FROM ERROR CORRECTING CODES

Cunsheng Ding, Xiaojian Tian, Xuesong Wang

Department of Computer Science, The Hong Kong University of Science and Technology, Clear Water Bay, Kowloon, Hong Kong, China

{cding,xjtian,cswxs}@cs.ust.hk

Abstract   In this paper, we present a simple and generic construction of systematic authentication codes which are optimal with respect to several bounds. The construction is based on error correcting codes. The authentication codes provide the best level of security with respect to spoofing attacks of various orders, including the impersonation and substitution attacks. The encoding of source states and the authentication verification are very simple and are perhaps the most efficient among all authentication systems.

Keywords: authentication codes, cryptography, linear codes.

1.   Introduction

Nowadays authentication and secrecy of messages are two basic security requirements in many computer and communication systems, and therefore two important areas in cryptography. Authentication codes are designed to provide sender and message authentication, and dates back to 1994 when Gilbert, MacWilliams and Sloane published the first paper in this area [see Gilbert, MacWilliams, Sloane, 1974]. Later Simmons [Simmos, 1984] developed a theory of unconditional authentication, which is analogous to Shannon's theory of unconditional secrecy [Shannon, 1949]. During the last twenty years codes that provide authentication and/or secrecy have been considered, and bounds and characterizations of these codes have been established, see, for example, [Gilbert, MacWilliams, Sloane, 1974], [Stinson 1990], [Casse, Martin, and Wild, 1998].

Most existing optimal authentication codes are constructed from combinatorial designs, and seem hard to implement. Even if some of them can be implemented in software or hardware, the implementation may not be efficient. In addition, these authentication codes provide protection against the impersonation and substitution attacks, but may not provide protection against spoofing attacks of order more than 1.

The purpose of this paper is to present a simple and generic construction of systematic authentication codes with the following properties:

- The authentication codes are optimal with respect to certain bounds.

- They offer the best security with respect to not only impersonation and substitution atacks, but also spoofing attacks of higher orders.

- The encoding of source states and authentication are extremely efficient and can be easily implemented in both software and hardware.

The construction of authentication codes presented here is based on error correcting codes, and is different from other constructions of authentication codes, see [Bierauer 1997], [Bierbrauer, Johansson, Kabatianskii and Smeets 1993], [Gilbert, MacWilliams, Sloane, 1974], [Kabatianskii, Smeets, and Johansson, 1996], [Simmons 1984], [Safavi-Naini and Seberry 1991], [Safavi-Naini, Wang and Xing 2001], using error correcting codes, in the sense that error correcting codes are employed to construct only the source states here in this paper.

2.   Systematic authentication codes and some bounds

A *systematic authentication code* is a three-tuple $\mathcal{A} = (\mathcal{S}, \mathcal{T}, \mathcal{E})$, where $\mathcal{S}$ is a set of *source states* and is associated with a probability distribution, $\mathcal{T}$ is a set of *authenticators* or *tags*, $\mathcal{E}$ is a set of *mappings* from $\mathcal{S}$ to $\mathcal{T}$, and is associated with a probability distribution. Each $e \in \mathcal{E}$ defines a one-to-one mapping

$$s \mapsto (s, e(s)),$$

which is called an *encoding rule*. Hence we also call $\mathcal{E}$ the *encoding rule space*. The set

$$\mathcal{M} = \mathcal{S} \times \{e(s) : e \in \mathcal{E} \text{ and } s \in \mathcal{S}\}$$

is called the *message space*. A systematic authentication code is used as follows.

In the basic model for authentication developed by Simmons [Simmons 1984], a *transmitter* communicates a sequence of distinct *source states* $s_1 s_2 \ldots s_i$ from the source state space $\mathcal{S}$ to a *receiver* by encoding them using one secret mapping $\{e \in \mathcal{E}\}$ into

$$[s_j, e(s_j)], \quad j = 1, 2, \ldots, i,$$

and sends this sequence of messages to the receiver. For each received message $[s, t]$, the receiver will compute $e(s)$ and check whether $e(s)$ matches t. If yes, the receiver will accept it as authentic, otherwise he/she will reject it.

The messages are sent to the receiver through a communication channel which may not be secure. A third party, an *opponent*, is involved in this model.

We assume that the opponent can intercept the messages transmitted and modify or replace them. Assume that the opponent has observed a sequence of messages

$$[s_j, e(s_j)], \quad j = 1, 2, \cdots, i,$$

sent from the transmitter to the receiver, where all s_j are pairwise distinct. The opponent now may not be able to determine the secret encoding rule e, but may have obtained partial information on e. The opponent then constructs another message $[s, t]$ such that $\Pr(e(s) = t)$ is maximal, and sends it to the receiver. If $e(s)$ indeed matches t, the opponent will be successful in attacking the system. This is called the *spoofing attack* of order i. The cases $i = 0$ and $i = 1$ are called *impersonation* and *substitution* attacks respectively. We use P_{d_i} to denote the opponent's maximum probability of success with respect to the spoofing attack of order i. We assume that the opponent knows the whole system except the secret encoding rule e shared by the transmitter and receiver.

It was proved in [Rosenbaum 1993] and [Sgarro 1993] that

$$P_{d_l} \geq 2^{H(\mathcal{E}|\mathcal{M}^{l+1}) - H(\mathcal{E}|\mathcal{M}^l)} \tag{1}$$

for any $l \geq 0$. Here $\mathcal{M}^l$ denotes the set of all possible sequences of l pairwise distinct messages and H denotes the uncertainty. This is the information-theoretic bounds which hold for authentication codes with and without secrecy. We shall need these bounds later.

We have also the following bounds.

Lemma 2.1. [Sinson 1990] *In any systematic authentication code, $P_{d_i} \geq \frac{|S|}{|\mathcal{M}|}$ for any $i \geq 0$.*

Lemma 2.2. [Sinson 1990] *In any systematic authentication code, if $P_{d_i} = \frac{|S|}{|\mathcal{M}|}$ for any $L \geq i \geq 0$, then*

$$|\mathcal{E}| \geq \left(\frac{|\mathcal{M}|}{|S|}\right)^{L+1}. \tag{2}$$

A *transversal design* $\mathrm{TD}_\lambda(t, \ell, n)$ is a triple (X, P, A), where X is a set of ℓn points, P is a partition of X into ℓ *groups* of n points each, and A is a set of λn^t blocks, each of which meets each group in a point, such that every t-subset of points from distinct groups occurs in exactly λ blocks.

A systematic authentication code for which $P_{d_i} = \frac{|S|}{|\mathcal{M}|}$ for $0 \leq i \leq L$, and in which

$$|\mathcal{E}| = \left(\frac{|\mathcal{M}|}{|S|}\right)^{L+1},$$

is said to be optimal with respect to the bound of (2). Such optimal authentication codes can be used to construct transversal designs.

Theorem 2.3. [Sinson 1990] *Assume there is an optimal systematic authentication code for ℓ source states, having v messages and $(v/\ell)^t$ encoding rules, and for which $P_{d_i} = \ell/v$ for $0 \leq i \leq t - 1$. Then there exists a transversal design* $\text{TD}_1(t, \ell, n)$, *where $n = v/\ell$.*

The following is a combinatorial bound.

Lemma 2.4. [Sinson 1992] *In any systematic authentication code, if $P_{d_i} = \frac{1}{|\mathcal{T}|}$ for $i = 0, 1$, then*

$$|\mathcal{E}| \geq |\mathcal{S}|(|\mathcal{T}| - 1) + 1. \tag{3}$$

This bound has been generalized into the following:

Lemma 2.5. [Kurosawa, Okada, Saido, and Stinson, 1998] *In any systematic authentication code, if $P_{d_i} = \frac{1}{|\mathcal{T}|}$ for $i = 0, 1, \cdots, t$, then*

$$|\mathcal{E}| \geq \begin{cases} \sum_{i=0}^{(t+1)/2} \binom{|\mathcal{S}|}{i}(|\mathcal{T}| - 1)^i & \text{if } t \text{ old} \\ \sum_{i=0}^{t/2} \binom{|\mathcal{S}|}{i}(|\mathcal{T}| - 1)^i + \binom{|\mathcal{S}|-1}{t/2}(|\mathcal{T}| - 1)^{t/2+1} & \text{if } t \text{ even} \end{cases} \tag{4}$$

The equality can be obtained if and only if $\mathcal{E}$ is uniformly distributed and the authentication matrix is an orthogonal array with strength $t + 1$.

This bound is a generalization of the classical Rao bound for orthogonal arrays [Colbourn and Dinitz 1996, p.180]. It is also an analogue of the sphere packing bound for linear codes [MacWilliams and Sloane 1977].

3. The construction of the authentication codes

We shall use $\mathbf{F}_q$ to denote the finite field with q elements. Let k be a positive integer, and let Δ be another integer such that $1 \leq \Delta \leq k$.

1. The source state space $\mathcal{S}$ of our authentication codes is a set of ℓ nonzero vectors in $\mathbf{F}_q^k$ such that any Δ of them are linearly independent. We assume that all source states are used equally likely. We will deal with the specific constructions of $\mathcal{S}$ later in Section 4.

2. The tag space $\mathcal{T}$ of our codes is $\mathbf{F}_q$.

3. Our space $\mathcal{E}$ is defined by $\mathcal{E} = \mathbf{F}_q^k$. Here all encoding rules are used equally likely, and for each $e \in \mathcal{E}$, the *encoding rule* defined by e is

$$u \mapsto (u, e(u)),$$

where $e(u) = e_1 u_1 + e_2 u_2 + \cdots + e_k u_k$ and $e = (e_1, e_2, \cdots, e_k) \in \mathcal{E}$ and $u = (u_1, u_2, \cdots, u_k) \in \mathcal{S}$.

Having described the construction of the authentication codes, we now calculate the deception probabilities P_{d_i} with respect to the spoofing attacks of different orders i.

Theorem 3.1. *For the authentication code described above, we have $P_{d_i} = \frac{1}{q}$, where $0 \leq i < \Delta \leq k$. In addition, if $\Delta + 1$ elements in S are linearly dependent, then $P_{d_\Delta} = 1$.*

Proof: Consider the spoofing attack of order i with $0 \leq i < \Delta$. In this case, the opponent has observed i pairwise distinct messages $[s_j, e(s_j)]$. Note that all encoding rules are used equally likely, and all source states are used with equal probability. Recall that by the definition of the source state space, the vectors $s_1, s_2, \cdots, s_i$ are linearly independent. Also the vectors $s_1, s_2, \cdots, s_i$ and s_{i+1} are linearly independent for any $s_{i+1} \in S$ which is different from $s_1, s_2, \cdots, s_i$. We have

$$P_{d_i} = \max_{s_h \in S, s_h \neq s_j, h \neq j} \max_{t_h \in F_q} \frac{|\{e : e(s_h) = t_h, h = 1, 2, \cdots, i + 1\}|}{|\{e : e(s_h) = t_h, h = 1, 2, \cdots, i\}|} = \frac{1}{q}.$$

Clearly, $P_{d_\Delta} = 1$ if $\Delta + 1$ elements in S are linearly dependent. $\square$

We now consider the optimality of the systematic authentication codes constructed in this section.

Theorem 3.2. *The authentication codes constructed above meet the information theoretic bounds of (1) and the bounds of Lemma 2.1 for all i with $0 \leq i < \Delta$. Hence they are optimal with respect to these bounds.*

Proof: The proofs are straightforward and are omitted here. $\square$

Theorem 3.3. *If $\Delta = k$, the authentication codes constructed above meet the bound of (2), and give transversal designs $TD_1(k, |S|, q)$.*

Proof: Note that the message space $\mathcal{M} = S \times F_q$. We have $\frac{|\mathcal{M}|}{|S|} = q$. Hence $|\mathcal{E}| = \left(\frac{|\mathcal{M}|}{|S|}\right)^\Delta$ if $\Delta = k$. Therefore the codes are optimal with respect to the bound of (2) if $\Delta = k$. The conclusion about transversal designs then follows from Theorem 2.3. $\square$

4. Specific constructions of authentication codes from error correcting codes

The construction of authentication codes presented in the previous section is generic. Different constructions of the source state space S yield different

systematic authentication codes without secrecy. In this section we present a
number of constructions of the source state space S using error correcting codes,
and thus obtain several classes of authentication codes within the framework of
the generic construction of Section 3.

In the generic construction, we require that $1 \leq \Delta \leq k$. In fact the case
$\Delta = 1$ is not interesting. So we shall consider only the cases that $2 \leq \Delta \leq k$.
In our construction, the source state space S is a subset S of $\mathbf{F}_q^k$ such that any
Δ of them are linearly independent. For optimality purpose and in view of the
bounds of (2) and (4), we wish to have the size of S maximal when Δ and
k are fixed. We are interested in not only the maximal size, but also specific
constructions of such sets.

Note that $\Delta \geq 2$. Our source state space S is actually a subset of the projective space $PG(k - 1, q)$. We now present a generic coding-theory construction
of the source state space S. Before doing this, we need some notations and
notions from finite geometries.

Let $PG(K, q)$ be the projective space of K dimensions over the finite field $\mathbf{F}_q$,
$q = p^h$ with p being prime, and let $|PG(K, q)| = \theta_K = (q^{K+1} - 1) / (q - 1)$.
A set of points in $PG(K, q)$ are linearly independent if and only if the vectors
representing them are a set of linearly independent vectors in the space $\mathbf{F}_q^{K+1}$.
In $PG(K, q)$, subspaces will be denoted by Π_i, where i is the dimension of the
subspace. A Π_0 is a *point*, a Π_1 is a *line*, and a Π_2 is a *plane*, a Π_3 is a *solid*,
and a Π_{K-1} is a *hyperplane* or *prime*.

An (ℓ, r)-*set* is a set of ℓ points at most r of which lie in Π_{r-1} but some $r + 2$
lie in a Π_r; that is, $r + 1$ points are always linearly independent, but some $r + 2$
points are linearly dependent. We use $M_r(k, q)$ to denote the maximum ℓ such
that an (ℓ, r)-set exists.

Any (ℓ, r)-set could be used as the source state space S in the construction of
Section 3, and the authentication code obtained will still be optimal with respect
to the information-theoretic lower bounds of (1). But we are interested more in
the lower bound of (2), and thus the case that $\Delta = k$. The construction of (ℓ, r)-
sets is an important area in finite geometries [Hirschfeld and Storme 1998].
Here we intend to present some (ℓ, r)-sets constructed from error correcting
codes and thus some constructions of authentication codes within the generic
construction of Section 3.

Linear error correcting codes can also be used to construct (ℓ, r)-sets. An
$[\ell, k]$ linear code C over $\mathbf{F}_q$ is a k-dimensional subspace of $\mathbf{F}_q^\ell$. A *generator
matrix* G of C is any matrix whose row vectors form a basis of the subspace C.
The column vectors of any generator matrix G of an $[\ell, k]$ linear code form an
(ℓ, r)-set, where $r = d - 2$ and d is the minimum distance of the dual code of
C. Of course the size of such an (ℓ, r)-set may not be maximal, i.e., it may be
smaller than $M_r(k, q)$. However, such an (ℓ, r)-set can be used as the source
state space in the construction of Section 3, and the authentication code is still

optimal with respect to the information theoretic bounds of (1) and the bounds of Lemma 2.1, but may or may not be optimal with respect to the bound of (2).

The coding theory construction above of (ℓ, r)-sets is generic and effective. As long as the minimum distance of a linear code over $\mathbf{F}_q$ is known, any generator matrix of its dual code gives an (ℓ, r)-set. In the following subsections, we use specific error correcting codes to construct (ℓ, r)-sets, and thus the source state space S and the corresponding authentication codes.

4.1   The construction from MDS codes

An $[\ell, k, d]$ linear code with $d = \ell - k + 1$ is called *maximum distance separable*, or MDS for short. The dual of an $[\ell, k, d]$ MDS code is also an $[\ell, \ell - k, k + 1]$ MDS code. Let C be an $[\ell, k, d]$ MDS code over $\mathbf{F}_q$, and G be a generator matrix of C. Define S to be the set of column vectors of G. Since $C^{\perp}$ has minimum distance $k + 1$, any k elements of S are linearly independent and at least one set of $k + 1$ elements of S are linearly dependent. With this source state space S, the generic construction of Section 3 gives an authentication code $(S, \mathcal{T}, \mathcal{E})$ with

$$|S| = \ell, \ |\mathcal{T}| = q, \ |\mathcal{E}| = q^k, P_{d_i} = \frac{1}{q}$$

for $i = 0, 1, \cdots, k - 1$, but $P_{d_k} = 1$. This authentication code is optimal with respect to the bounds of (1) and Lemma 2.1.

MDS codes over $\mathbf{F}_q$ with the following parameters exist [MacWilliams and Sloane, Chap. 11]:

1. $[q - 1, k, q - k]$ Reed-Solomon codes, where $1 \le k \le q - 1$;

2. $[q, k, q - k + 1]$ extended RS codes, where $1 \le k \le q$;

3. $[q + 1, k, q - k + 2]$ cyclic codes, where $1 \le k \le q + 1$.

Consider the authentication code based on a $[q + 1, k, q - k + 2]$ code, where q is odd. If $k = 2$, it is easily checked that the bound of (4) is met. If $k = 3$ or $k = 4$, this bound is not met.

A *permutation polynomial* $F(x)$ over $\mathbf{F}_q$ is a polynomial over $\mathbf{F}_q$ and a permutation of $\mathbf{F}_q$. An *o-polynomial* is a permutation polynomial over $\mathbf{F}_q$ of degree at most $q - 2$, satisfying $F(0) = 0$ and $F(1) = 1$, and such that $F_s(x) = (F(x + s) - F(s))/x$ is a permutation polynomial for each $s \in \mathbf{F}_q$, satisfying $F_s(0) = 0$. There are several classes of o-polynomials over $\mathbf{F}_q$ when q is even [Hirschfeld and Storme 1996]. For example, $F(x) = x^2$ is an o-polynomial.

Let $t_0, t_1, \cdots, t_{q-1}$ denote all the elements of $\mathbf{F}_q$, and let $F(x)$ be an o-polynomial over $\mathbf{F}_q$. If q is even, the following matrix

$$
G = \begin{bmatrix}
0 & 0 & 1 & 1 & \cdots & 1 \\
0 & 1 & t_0 & t_1 & \cdots & t_{q-1} \\
1 & 0 & F(t_0) & F(t_1) & \cdots & F(t_{q-1})
\end{bmatrix}
$$

generates a $[q + 2, 3, q]$ MDS code. Its dual has parameters $[q + 2, q - 1, 4]$. Since there are a number of o-polynomials $F(x)$, we have a number of classes of $[q + 2, 3, q]$ MDS codes over $\mathbf{F}_q$. We now prove that the authentication codes based on these MDS codes are also optimal with respect to the bound of (4).

In this case, we have $t = 2$, $|\mathcal{S}| = q + 2$ and $|\mathcal{T}| = q$. Hence

$$
\sum_{i=0}^{t/2} \binom{|\mathcal{S}|}{i} (|\mathcal{T}| - 1)^i + \binom{|\mathcal{S}| - 1}{t/2} (|\mathcal{T}| - 1)^{t/2+1} = q^3 = |\mathcal{E}|.
$$

Thus the bound of (4) is met.

4.2   The construction from the duals of almost MDS codes

An $[\ell, k, d]$ linear code with $d = \ell - k$ is called *almost maximum distance separable*, or almost MDS for short [De Boer 1996]. Let C be an $[\ell, \ell - k, k]$ almost MDS code over $\mathbf{F}_q$. Then $C^{\perp}$ has dimension k. Define $\mathcal{S}$ to be the set of column vectors of a generator matrix of $C^{\perp}$. Then any $k - 1$ elements of $\mathcal{S}$ are linearly independent, but some k elements of $\mathcal{S}$ are linearly dependent. With this source state space $\mathcal{S}$, the generic construction of Section 3 gives an authentication code $(\mathcal{S}, \mathcal{T}, \mathcal{E})$ with

$$
|\mathcal{S}| = \ell, \ |\mathcal{T}| = q, \ |\mathcal{E}| = q^k, \ P_{d_i} = \frac{1}{q}
$$

for $i = 0, 1, \cdots, k - 2$, but $P_{d_{k-1}} = 1$. This authentication code is optimal with respect to the first two types of bounds described in Section 2, but not optimal with respect to the bound of (2) in general. However it could be optimal with respect to this bound in certain cases, as demonstrated below.

The dual of a MDS code is still MDS. But the dual of an almost MDS code may not be almost MDS. If a code and its dual both are almost MDS (AMDS), the code is called *near MDS* (in short, NMDS). In fact almost MDS codes are much more complicated than MDS codes. For example, the weight distribution of MDS codes is known, but that of almost MDS codes is not determined.

There are several classes of almost MDS codes. The $[q^2 + q + 1, q^2 + q - 2, 3]$ Hamming code is almost MDS. Its dual is the $[q^2 + q + 1, 3, q^2]$ simplex code.

Theorem 4.1. *The authentication code based on the dual of the $[q^2+q+1, 3, q^2]$ simplex code has*

$$|\mathcal{S}| = q^2 + q + 1, \ |\mathcal{T}| = q, \ |\mathcal{E}| = q^3, \ P_{d_i} = \frac{1}{q}$$

for $i = 0, 1$, but $p_{d_2} = 1$.

This authentication code is optimal with respect to the first two types of bounds described in Section 2, but not optimal with respect to the bound of (2). However it is also optimal with respect to the bound of (4).

Theorem 4.2. *There is a $[q^2 + 1, q^2 - 3, 4]$ almost MDS code C over $\mathbf{F}_q$ for odd q [Hirschfeld and Storme 1996]. The authentication code based on $C^\perp$ has*

$$|\mathcal{S}| = q^2 + 1, \ |\mathcal{T}| = q, \ |\mathcal{E}| = q^4, \ P_{d_i} = \frac{1}{q}$$

for $i = 0, 1, 2$, but $p_{d_3} = 1$.

This authentication code is optimal with respect to the first two types of bounds described in Section 2, but not optimal with respect to the bounds of (2) and (4).

4.3   The construction from the duals of perfect codes

An $[\ell, k, 2e + 1]$ code over $\mathbf{F}_q$ is called *perfect* if

$$q^{\ell-k} = \sum_{i=0}^{e} \binom{\ell}{i}(q-1)^i. \tag{5}$$

Theorem 4.3. *Let C be an $[\ell, \ell - k, 2e + 1]$ code over $\mathbf{F}_q$. The authentication code based on $C^\perp$ has the following parameters*

$$|\mathcal{S}| = \ell, \ |\mathcal{T}| = q, \ |\mathcal{E}| = q^k, \ P_{d_i} = \frac{1}{q}$$

for $i = 0, 1, \cdots, 2e - 1$.

This authentication code is optimal with respect to the bound of (2) if and only if $k = 2e$, i.e., if and only if C is an $[\ell, \ell - k, k + 1]$ MDS code.

This authentication code is optimal with respect to the bound of (4) if and only if C is perfect.

Proof: We now prove that the authentication code is optimal with respect to the bound of (2) if and only if $k = 2e$, i.e., if and only if C is an $[\ell, \ell - k, k + 1]$ MDS code. Note that $|\mathcal{E}| = q^k$, $|\mathcal{M}|/|\mathcal{S}| = q$, and $L = 2e - 1$. We see that the bound of (2) is met if and only if $k = 2e$, which is equivalent to C being an $[\ell, \ell - k, k + 1]$ MDS code.

The bound of (4) holds if and only if $|\mathcal{E}| = \sum_{i=0}^{e} \binom{|\mathcal{S}|}{i}(|\mathcal{T}| - 1)^i$ which becomes $q^k = \sum_{i=0}^{e} \binom{\ell}{i}(q-1)^i$. This is equivalent to $q^\ell = q^{\ell-k} \sum_{i=0}^{e} \binom{\ell}{i}(q-1)^i$ which is the condition for $\mathcal{C}$ being perfect.   □

There are only three types of perfect codes [MacWilliams and Sloane, Chapter 6, Section 10]:

- The $[(q^k - 1)/(q - 1), (q^k - 1)/(q - 1) - k, 3]$ Hamming codes.

- The binary $[23, 12, 7]$ Golay code.

- The ternary $[11, 6, 5]$ Golay code.

The column vectors of the generator matrix of the simplex code consists of all points in $\mathrm{PG}(k - 1, q)$. The simplex code has length $(q^k - 1)/(q - 1)$, dimension k, and minimum distance q^{k-1}. Its dual code is the $[(q^k - 1)/(q - 1), (q^k - 1)/(q - 1) - k, 3]$ Hamming code. These three perfect codes give authentication codes which are optimal with respect to the bound of (4).

5.   Open problems

The bound of (4) is not tight at least in the case t being odd. It may be strengthened into a bound similar to the Johnson bound using similar techniques [MacWilliams and Sloane, pp. 532–533].

Open problem: Strengthen the bound of (4).

With respect to our construction and the bound of (4), perfect codes give the best authentication codes. However, Hamming codes are the only class of perfect codes which contains infinitely many codes. We may use nearly perfect and quasi-perfect codes [Pless 1998, p. 132] to construct good authentication codes within the framework of the generic construction described in this paper.

6.   Concluding remarks

In this paper we presented a simple and generic construction of systematic authentication codes with the properties outlined in Section 1. The encoding of a source state or the authentication checking takes only $k - 1$ additions and k multiplications over $\mathbf{F}_q$, and thus are extremely efficient. In addition, the systematic authentication codes constructed in this paper are based on error correcting codes and thus are algebraic.

References

[1]   J. Bierbrauer. Universal hashing and geometric codes. *Designs, Codes and Cryptography*, 11:207–221, 1997.

[2] J. Bierbrauer, T. Johansson, G. Kabatianskii, and B. Smeets. On families of hash functions via geometric codes and concatenation. In *Advances in Cryptology - Crypto' 93*, pages 331–342, 1994. Springer LNCS 773.

[3] L.R.A. Casse, K.M. Martin, and P.R. Wild. Bounds and characterizations of authentication/secrecy schemes. *Designs, Codes and Cryptography*, 13:107–129, 1998.

[4] C.J. Colbourn and J.H. Dinitz. CRC Handbook of Combinatorial Designs. CRC Press, 1996.

[5] M.A. De Boer. Almost MDS codes. *Designs, Codes and Cryptography*, 9:143–155, 1996.

[6] E.N. Gilbert, F.J. MacWilliams, and N.J.A. Sloane. Codes which detect deception, *Bell System Tech. Journal*, 53:405–424, 1974.

[7] J.W.P. Hirschfeld, and L. Storme. The packing problem in statistics, coding theory and finite projective spaces. *J. Statistics Planning and Inference*, 72:355–380, 1998.

[8] G.A. Kabatianskii, B. Smeets, and T. Johansson. On the cardinality of systematic authentication codes via error-correcting codes. *IEEE Trans. Inform. Theory*, 42:566–578, 1996.

[9] K. Kurosawa, K. Okada, H. Saido, and D.R. Stinson. New combinatorial bounds for authentication codes and key predistribution schemes. *Designs, Codes and Cryptography*, 15:87–100, 1998.

[10] J.L. Massey. Cryptography: A selective survey. In *Digital Communications*, pages 3–21. North-Holland, 1986.

[11] F.J. MacWilliams and N.J.A. Sloane. *The theory of Error Correcting Codes*. North-Holland, Amsterdam, 1977.

[12] V.S. Pless, An introduction to algebraic codes. In *Handbook of Coding Theory I*, pages 1–139. North-Holland/Elsevier, 1998.

[13] R.S. Rees and D.R. Stinson. Combinatorial characterizations of authentication codes. *Designs, Codes and Cryptography*, 7:239–259, 1996.

[14] U. Rosenbaum. A lower bound on authentication after having observed a sequence of messages. *J. Cryptology*, 6:135–156, 1993.

[15] R. Safavi-Naini, and J. Seberry. Error correcting codes for authentication and sublimal channels. *IEEE Trans. Information Theory*, 37(1):13–17, 1991.

[16] R. Safavi-Naini, H. Wang, and C. Xing. Linear authentication codes: bounds and constructions. In *Progress in Cryptology – INDOCRYPT 2001*, pages 127–135. Springer LNCS 2247, 2001.

[17] A. Sgarro. Information-theoretic bounds for authentication frauds. *J. Computer Security*, 2:53–63, 1993.

[18] C.E. Shannon. Communication theory of secrecy systems. *Bell System Tech. Journal*, 28:656–715, 1949.

[19] G.J. Simmons. Authentication theory/coding theory. In *Advances in Cryptology: Proceedings of Crypto' 84*, pages 411–432. Springer LNCS 196, Berlin, 1985.

[20] D.R. Stinson. The combinatorics of authentication and secrecy codes. *J. Cryptology*, 2:23–49, 1990.

[21] D.R. Stinson. Combinatorial characterizations of authentication codes. *Designs, Codes, and Cryptography*, 2:175–187, 1992.

ON COEFFICIENTS OF BINARY EXPRESSION OF INTEGER SUMS*

Bao Li, Zongduo Dai
State Key Laboratory of Information Security
Graduate School, Chinese Academy of Sciences
100039 Beijing, China
lb@is.ac.cn

Abstract   Binary expression of integer sums is investigated. A precise formula is presented for the coefficients of binary expressions of integer sums. This formula is a basis for some other results.

Keywords: integer, sequence, binary, ring

1.   Introduction

Sequences over $\mathbb{F}_2$ are very important both in theory and practice. They attract more and more attentions with the development of computer and modern cryptography. Sequences over $\mathbb{F}_2$ can be derived from sequences over $\mathbb{Z}_{2^e}$–the ring of integer ring modulo 2^e. Some way of deriving pseudorandom binary sequences from sequences over $\mathbb{Z}_{2^e}$ is investigated in [1]. And in the same paper the periods and linear complexities of the derived sequences are investigated as well.

Since the way of handling overflows in $\mathbb{Z}_{2^e}$ is different from that in $\mathbb{F}_{2^e}$, sequences over $\mathbb{Z}_{2^e}$ are of particular interest from an application point of view as they can be generated very efficiently on microprocessors when e is the word length of the processor.

In China cryptography researchers have noticed the significance of sequences over $\mathbb{Z}_{2^e}$ in 1980's and worked on them systematically [2-8].

Recently, studies on sequences over $\mathbb{Z}_{2^e}$ are still active [9-11]. As a natural generalization sequences over Galois rings have been discussed[12-14].

In the study of sequences over $\mathbb{Z}_{2^e}$ the carries of sequences over $\mathbb{Z}_{2^e}$ play an important role. We observe that the carries of sequences over $\mathbb{Z}_{2^e}$ are based

*Partial funding provided by State Key Laboratory of Information Security.

on properties of integer sums in $\mathbb{Z}_{2^e}$. Therefore, we focus on study of integer sums in this paper and present a precise formula for the coefficients of binary expressions of integer sum utilizing Lucas' Theorem. We point out that this formula can be used to improve the results in [1] as an application . And as another application we plan to use this formula to present a proof for the result given in [5] which has not been given a complete proof until now. We point out that our result can be generalized over $\mathbb{Z}_{p^e}$ further and may cause generalization of concepts and results of sequences over $\mathbb{Z}_{2^e}$ to that over $\mathbb{Z}_{p^e}$ where p is a prime.

2.   preparation

At first we present Lucas' Theorem and its proof for the sake of convenience.

Lemma 2.1 (Lucas' Theorem[15]**).** *Let* $a = \sum_{i=1}^{N} a_i 2^i$, $a_i \in \{0,1\}$, $b = \sum_{i=1}^{N} b_i 2^i$, $b_i \in \{0,1\}$. *Then*

$$\binom{a}{b} = \prod_{i=1}^{N} \binom{a_i}{b_i} \quad (\text{mod } 2). \tag{1}$$

Proof. Let x be an indeterminate and consider the polynomial $(x+1)^a \in \mathbb{F}_2[x]$ where $\mathbb{F}_2$ the binary field. Expanding $(x+1)^a$ over $\mathbb{F}_2$ we have that

$$(x + 1)^a = \sum_{k=0}^{a} \binom{a}{k} x^k. \tag{2}$$

On the other hand,

$$(x + 1)^a = (x+1)^{\sum_{i=1}^{N} a_i 2^i} = \prod_{i=1}^{N} (x^{2^i} + 1)^{a_i} = \prod_{i=1}^{N} \left(\binom{a_i}{1} x^{2^i} + 1 \right). \tag{3}$$

Consider the coefficient of x^b in (3). For the uniqueness of binary expression of b it is easy to see that the efficient of x^b is

$$\prod_{b_i=1} \binom{a_i}{1} = \prod_{i=1}^{N} \binom{a_i}{b_i}. \tag{4}$$

Taking into account of (2), (3) and (4) one achieves (1)   □

Secondly, we introduce a classical result on the binary expression of bit sums which was presented in [16]. As an application of Lucas' Theorem we present the proof of this result.

Lemma 2.2. [16] *If $c_i \in \{0,1\}, 1 \leq i \leq N$, and $a_j \in \{0,1\}, 1 \leq j < e$, and*

$$\sum_{i=1}^{N} c_i = \sum_{j \geq 0} a_j 2^j, \tag{5}$$

then

$$a_j = \sum_{1 \leq i_1 < i_2 < \cdots < i_{2^j} \leq N} c_{i_1} c_{i_2} \cdots c_{i_{2^j}} \quad (\mathrm{mod}\ 2). \tag{6}$$

Proof. Firstly we assume $c_i \neq 0$ for all $1 \leq i \leq N$. In this case we have

$$\binom{N}{2^j} = \sum_{1 \leq i_1 < i_2 < \cdots < i_{2^j} \leq N} c_{i_1} c_{i_2} \cdots c_{i_{2^j}}. \tag{7}$$

And by Lemma 2.1 we have

$$\binom{N}{2^j} = \prod_{i=1}^{N} \binom{a_i}{b_i} = a_j \quad (\mathrm{mod}\ 2), \qquad \text{where } b_i = \begin{cases} 1, & i = j, \\ 0, & i \neq j. \end{cases} \tag{8}$$

Thus, combining (7) and (8) we obtain equation (6).

As for the case of $c_i = 0$ for some i we may assume that

$$\sum_{i=1}^{N} c_i = n.$$

And it is easy to see that equations (7) and (8) are transformed into

$$\binom{n}{2^i} = \sum_{1 \leq i_1 < i_2 < \cdots < i_{2^j} \leq N} c_{i_1} c_{i_2} \cdots c_{i_{2^j}}; \tag{9}$$

and

$$\binom{n}{2^i} = \prod_{i=1}^{N} \binom{a_i}{b_i} = a_j \quad (\mathrm{mod}\ 2), \qquad \text{where } b_i = \begin{cases} 1, & i = j, \\ 0, & i \neq j. \end{cases} \tag{10}$$

Combining (9) and (10) we obtain equation (6) as well. $\square$

3.   Main theorem

Suppose there are N integers which are expressed in binary forms. If the sum of them are expressed in binary form, we want to know the coefficients of the binary expression as a combination of the integers binary coefficients. We present a precise formula in following which can be view as a generalization of Lemma 2.2.

Firstly, we need some notations and concepts to make the description of the result clear and easy.

For simplicity of narrative we suppose the integers involved are all in the ring $\mathbb{Z}_{2^{e+1}}$–integer ring modulo 2^{e+1} and expressed in binary forms.

Definition 3.1. *Set*

$$\Omega_k = \{(i,k)|0 \leq i \leq N\} \tag{11}$$

and

$$\Omega = \bigcup_{0 \leq k \leq e} \Omega_k. \tag{12}$$

Let $x_{i,k}$ be indeterminate, $(i,k) \in \Omega_k$. By **X** we mean a set of indeterminates $x_{i,k}$, with $(i,k) \in \Omega_k, 0 \leq k \leq e$.

Let $I = \bigcup I_k, I_k \subseteq \Omega_k$, define

$$|I| = \sum_{0 \leq k \leq e} |I_k| 2^k, \tag{13}$$

where $|I_k|$ is the number of elements in I_k.

Definition 3.2.

$$\pi_I(\boldsymbol{X}) = \prod_{(i,k) \in I} x_{i,k} \tag{14}$$

and

$$f_\lambda(\boldsymbol{X}) = \sum_{I \subseteq \Omega, |I| = \lambda} \pi_I(\boldsymbol{X}) \tag{15}$$

are called minor term function *and* λ-function *respectively.*

We may consider now the integer sum. We assume the integers have been expressed in binary forms and present the result in terms of notations and concepts we defined above.

Theorem 3.3. *Suppose* $n_i = \sum_{0 \leq k \leq e} d_{i,k} 2^k, d_{i,k} \in \{0,1\}, 1 \leq i \leq N, 0 \leq k \leq e$ *are positive integers in* $\mathbb{Z}_{2^{e+1}}$. *Let*

$$d = \sum_{1 \leq i \leq N} n_i = \sum_{1 \leq i \leq N, 0 \leq k < e} d_{i,k} 2^k \pmod{2^e}. \tag{16}$$

If $d = \sum_{j \geq 0} D_j 2^j, D_j \in \{0,1\}$, *then*

$$D_j = f_{2^j}(\boldsymbol{X}) \mid_{x_{i,k} = d_{i,k}} \pmod{2}. \tag{17}$$

Proof. Suppose $\mathbf{d} = (d_{i,k})$. Define

$$\Omega_k(\mathbf{d}) = \{(i,k) \mid d_{i,k} = 1, 1 \le i \le N\} \subseteq \Omega_k, \qquad (18)$$

$$\Omega(\mathbf{d}) = \bigcup_{0 \le k \le e} \Omega_k(\mathbf{d})$$
$$= \{(i,k) \mid d_{i,k} = 1, 1 \le i \le N, 0 \le k \le e\} \subseteq \Omega. \qquad (19)$$

We prove this theorem by claiming several facts.

If for a given $I \subseteq \Omega$ we define

$$\pi_I(\mathbf{d}) = \pi_I(\mathbf{X}) \mid_{x_{i,k} = d_{i,k}}, \qquad (20)$$

then according to definition of $\Omega(\mathbf{d})$, the following fact is evident.

Fact 3.4.

$$\pi_I(\mathbf{d}) = \begin{cases} 1, & I \subseteq \Omega(\mathbf{d}), \\ 0, & otherwise. \end{cases} \qquad (21)$$

Fact 3.5. *Let $d_k = \sum_{(i,k) \in \Omega_k(\mathbf{d})} d_{i,k}$. Then*

$$\binom{d_k}{j_k} = \sum_{\substack{I_k \subseteq \Omega_k(\mathbf{d}) \\ |\bar{I}_k| = j_k}} \pi_{I_k}(\mathbf{d}) \quad (\text{mod } 2). \qquad (22)$$

Proof. This is another expression of Lemma 2.2 in our language. $\qquad \square$

Fact 3.6.

$$\binom{d_0}{j_0} \cdots \binom{d_e}{j_e} = \sum_{\substack{I = \bigcup_{0 \le k \le e} \subseteq \Omega(\mathbf{d}) \\ |\bar{I}_k| = j_k}} \pi_I(\mathbf{d}) \quad (\text{mod } 2). \qquad (23)$$

Proof. By Fact 3.5 we have

$$\binom{d_0}{j_0} \cdots \binom{d_e}{j_e} = \Big(\sum_{\substack{I \subseteq \Omega(\mathbf{d}) \\ |\bar{I}_0| = j_0}} \pi_{I_0}(\mathbf{d}) \Big) \cdots \Big(\sum_{\substack{I_k \subseteq \Omega_e(\mathbf{d}) \\ |\bar{I}_e| = j_e}} \pi_{I_e}(\mathbf{d}) \Big).$$

And if $I = \bigcup_{0 \le k \le e} I_k$, then $\pi_I(\mathbf{X}) = \pi_{I_0}(\mathbf{X}) \cdots \pi_{I_e}(\mathbf{X})$. So we have

$$(\sum_{\substack{I \subseteq \Omega(\mathbf{d}) \\ |\bar{I}_0| = j_0}} \pi_{I_0}(\mathbf{d})) \cdots (\sum_{\substack{I_k \subseteq \Omega_e(\mathbf{d}) \\ |\bar{I}_e| = j_e}} \pi_{I_e}(\mathbf{d}))$$

$$= \sum_{\substack{I = \bigcup_{0 \le k \le e} I_k \subseteq \Omega(\mathbf{d}) \\ |\bar{I}_k| = j_k}} \pi_I(\mathbf{d})$$

$$= \sum_{\substack{I = \bigcup_{0 \le k \le e} I_k \subseteq \Omega \\ |\bar{I}_k| = j_k}} \pi_I(\mathbf{d}) \quad (\text{mod } 2).$$

The last equality is from Fact 3.4. $\square$

From Fact 3.6 we may directly claim

Fact 3.7.

$$\sum_{\substack{(j_0, \cdots, j_e) \\ \sum_{0 \le k \le e} j_k 2^k = \lambda}} \binom{d_0}{j_0} \cdots \binom{d_e}{j_e} = \sum_{\substack{I \subseteq \Omega(\mathbf{d}) \\ |I| = \lambda}} \pi_I(\mathbf{d}) \quad (\text{mod } 2). \qquad (24)$$

Fact 3.8. *Suppose $d = \sum_{(i,k) \in \Omega} d_{i,k} 2^k$, then*

$$\binom{d}{\lambda} = f_\lambda(X) \mid_{X=d} \quad (\text{mod } 2), \qquad (25)$$

where $f_\lambda(X) = \sum_{I \subseteq \Omega, |I| = \lambda} \pi_I(X)$.

Proof. We expand the following polynomial over $\mathbb{F}_2$

$$(z+1)^d = \prod_k (z+1)^{d_k 2^k} = \prod_k (z^{2^k} + 1)^{d_k}$$

$$= \prod_{0 \le k \le e} \sum_{0 \le j_k \le d_k} \binom{d_k}{j_k} z^{2^k j_k} \qquad (26)$$

$$= \sum_{\lambda \ge 0} z^\lambda \sum_{\substack{(j_0, \cdots, j_e) \\ \sum j_k 2^k = \lambda}} \binom{d_0}{j_0} \cdots \binom{d_e}{j_e}.$$

On the other hand,

$$(z+1)^d = \sum_\lambda \binom{d}{\lambda} z^\lambda. \qquad (27)$$

Thus, comparing the coefficients of z^λ in (26) and (27) we have

$$
\begin{aligned}
\binom{d}{\lambda} &= \sum_{\substack{(j_0,\cdots,j_e) \\ \sum j_k 2^k = \lambda}} \binom{d_0}{j_0} \cdots \binom{d_e}{j_e} \\
&= \sum_{\substack{I \subseteq \Omega(\mathbf{d}) \\ |I|=\lambda}} \pi_I(\mathbf{d}) = \sum_{\substack{I \subseteq \Omega \\ |I|=\lambda}} \pi_I(\mathbf{X}) \mid_{\mathbf{X}=\mathbf{d}} \\
&= f_\lambda(\mathbf{X}) \mid_{\mathbf{X}=\mathbf{d}} \quad (\mathrm{mod}\ 2).
\end{aligned}
$$

□

By Lucas' Theorem (Lemma 2.1) taking $a = d$ and $b = 2^j$ we have $D_j = \binom{d}{2^j}$. By Fact 3.8 we have $\binom{d}{2^j} = f_{2^j}(\mathbf{X}) \mid_{\mathbf{X}=\mathbf{d}}$. This completes the proof of Theorem 3.3.

□

4.   Conclusions

In [1], the concept of carry sequences is defined for linear recurring sequences over $\mathbb{Z}_{2^e}$. Applying the result in this paper the concept and result in [1] can be simplified. For the reason of limited space and complicated notations the work will be left for another paper.

As a matter of fact, our original ambition is to present a proof for the main lemma in [5] and finally give the detailed proof of the lower bounds of linear complexity of sequences over $\mathbb{Z}_{2^e}$. In the process of working on this task, we gradually clarify the essence of the problem and abstract the key concepts and reach the result. Applying this result we can achieve our original goal, but it will be a lengthy work and we will not do it in this paper.

We point out that our result may be generalized without any difficulties in essence. The more significant goal in our mind is to generalize the concepts and results of sequences over $\mathbb{Z}_{p^e}$. We consider it a more difficult task to complete.

References

[1] Z. D. Dai, Binary sequences derived from ML-sequences over rings I: Periods and minimal polynomials, J. Cryptology, Vol. 5, No. 4, 193-207, 1992.

[2] K. C. Tzeng, etc., Maximal length sequences over rings, Reports in Data Center of Security, 1987.

[3] M. Q. Huang, Analysis and cryptologic evaluation of primitive sequences over an integer residue ring, Doctoral dissertation of Graduate School of USTC, Academia Sinica, 1988.

[4] Z. D. Dai and M. Q. Huang, A criterion for primitiveness of polynomials over $Z/(2^d)$, Chinese Science Bulletin, Vol. 36, No. 11, 892-895, June 1991.

[5] Z. D. Dai and T. Beth and D. Gollmann, Lower bounds for the linear complexity of sequences over residue rings, Advances in Cryptology-EUROCRYPT'90, Springer-Verlag, Editor: I. B. Damgard, 189-195, 1991.

[6] M. Q. Huang and Z. D. Dai, Projective maps of linear recurring sequnences with maximal p-adic periods, Fibonacci Quart 30, No. 2, 139-143, 1992.

[7] W. F. Qi and J. J. Zhou, Distribution of 0 and 1 in highest level of primitive sequences over $Z/(2^e)$, Science in China, Series A, 40(6), 606-611, 1997.

[8] W. F. Qi and J. H. Yang, J. J. Zhou, ML-Sequences over rings $Z/(2^e)$, Advances in Cryptology- ASIACRYPT¡¯98, LNCS., 1514, 315¨C326, Berlin Heidelberg: Springer-Verlag, 1998.

[9] Qi Wenfeng and Zhu Fengxiang, Distribution of 0 and 1 in Compression Sequences over $Z/(2^e)$, MATHEMATICA APPLICATA, Vol.13, No.1, 102-108, 2000.

[10] Qi Wenfeng and Zhu Xuanyong, Injectiveness of Compression Mappings on Primitive Sequences over Galois Rings, ACTA MATHEMATICA SINICA, Vol.44, No.3, 445-452, May, 2001.

[11] FAN Shuqin and HAN Wenbao, 0,1 distribution in the highest level sequences of primitive sequences over $Z/(2^e)$, Science in China Series A, Vol.46, No.4, 516-524, 2003.

[12] Kuz'min A. S., Nechaev A. A., Linear Recurring Sequences over Galois Ring, Russion Mathmatical Surveys, 48: 171¨C172, 1993.

[13] O. V. Kamlovskii and A. S. Kuz'min, Distribution of elements on cycles of linear recurrent sequences over Galois rings, Russ. Math. Surv., 53 (2), 392-393, 1998.

[14] O. V. Kamlovsky A. S. Kuzmin, Bounds for the number of occurrences of elements in a linear recurring sequence over a Galois ring, FUNDAMENTALNAYA I PRIKLADNAYA MATEMATIKA (FUNDAMENTAL AND APPLIED MATHEMATICS) Vol. 6, No. 4, 1083-1094, 2000.

[15] F. J. MacWilliam and N. J. A. Sloane, The Theory of Error-Correcting Codes, Norht-Holland Publishing Company, 1977.

[16] R. A. Rueppel, Analysis and design of stream ciphers, Springer-Verlag, Berlin, 1986.

[17] R. Lidl and H. Niederreiter, Finite Fields, Encyclopaedia of Mathmatics and Its Applications, Vol. 20, Addison-Wesley, Reading, MA, 1983.

[18] H. Beker and F. Piper, Cipher Systems, Northwood Books, London, 1982.

[19] M. Ward, The arithmetical theory of linear recurring series, Trans. Amer. Math. Soc., Vol. 35, 600-628, July 1933.

[20] N. Zierler and H. Mills, Products of linear recursive sequences, J. Algebra, Vol. 27, 147-157, 1973.

A NEW PUBLICLY VERIFIABLE PROXY SIGNCRYPTION SCHEME

Zhang Zhang, Qingkuan Dong

National Key Lab. of ISN, Xidian University, Xi'an, 710071, P. R. China
zzhang74@yahoo.com.cn

Mian Cai

School of Electronic Information and Control Engineering
Beijing University of Technology , 100022, P. R. China

Abstract   Proxy signcryption scheme is the combination of proxy signature and encryption. This paper presents a new proxy signcryption scheme based on Shin et al.'s DSA-verifiable signcryption Scheme. As compared to the previous schemes, the proposed scheme has the following advantages: it can protect the proxy signer against the original signer's forgery attack; it can be publicly verified; it provides forward secrecy property with respect to the proxy signer.

Keywords: signcryption, proxy signature, proxy signcryption

Introduction

Secure and authenticated message transmission is one of the major aims of computer and communication security research. The traditional method to achieve this is signature followed by encryption. Nyberg and Rueppel suggested an authenticated encryption scheme of this type as an application of their message recovery signature scheme [1]. An authenticated encryption scheme is a message transmission scheme which sends messages in a secure and authentic way. Basically, an authenticated encryption scheme should satisfy the following properties:

(1)confidentiality: it is computationally infeasible for an adaptive attacker to find out any secret information from a ciphertext.

(2) authenticity (unforgeability): it is computationally infeasible for an adaptive attacker to masquerade as the sender in sending a message.

(3) nonrepudiation: it is computationally feasible for a third party to settle a

dispute between the sender and the receiver in an event where the sender denies the fact that he is the originator of the message.

In 1997, Zheng introduced the concept of signcryption schemes [2]. It was claimed that authenticity, confidentiality and nonrepudiation were gained and the efficiency is superior to all schemes based on the aforementioned paradigms. However, Petersen et al. pointed out that the way to gain nonrepudiation violates the confidentiality in Zheng's scheme [3]. Later on many improved schemes are suggested [4, 5, 6].

In 1998, Gamage et al. introduced the notion of proxy signcryption [7]. Proxy signcryption scheme is the combination of proxy signature [8] and encryption. Since Gamage et al.'s proxy signcryption scheme is based on Zheng's original scheme, it has some weaknesses as Zheng's scheme. Later on, Jung et al. proposed a scheme to overcome the weakness in Gamage et al.' scheme [5]. However, Jung et al.'s scheme can not be publicly verified. In this paper, the authors propose a new proxy signcryption scheme based on Shin et al.'s DSA-verifiable signcryption Scheme. As compared to the previous schemes, the proposed scheme has the following advantages: it can protect the proxy signer against the original signer's forgery attack; it can be publicly verified; it has forward secrecy property.

1.   Related works

In this section, we give a brief description of Gamage et al.'s proxy signcryption scheme and Shin et al.'s DSA-verifiable signcryption Scheme. Throughout this paper, we will use the following setting. Let p, q, and g be the public parameters: p a large prime, q a large prime divisor of $p-1$ and g an element in Z_p of order q. Let H be a one-way hash function. Encrypting a message m with a key k is indicated by $E_k(m)$ while decrypting a cipher c with a key k is denoted by $D_k(c)$. We use $KH_k(m)$ denote hashing a message m with KH under a key k. We assume that the original signer is Alice and the receiver is Bob. The secret key and public key pairs of Alice and Bob are (x_A, y_A) and (x_B, y_B), where $y_A = g^{x_A} \bmod p$ and $y_B = g^{x_B} \bmod p$. The secret key and public key pairs of the proxy signer is (x_P, y_P), where $y_P = g^{x_P} \bmod p$.

1.1   Gamage et al.'s proxy signcryption scheme

Proxy key generation: Alice randomly selects $x' \in Z_q^*$ and computes

$$K' = g^{x'} \bmod p \text{ and } x_{ap} = x_a + x'K' \bmod q.$$

Alice sends (x_{ap}, K') through a secure channel to the proxy signer. The proxy signer accepts x_{ap} as a valid proxy signature key only if the following equation is holds:

$$g^{x_{ap}} = y_a K'^{K'} \bmod p.$$

Proxy signcryption: The proxy signer randomly chooses $x \in Z_q^*$ and computes

$$k = y_b^x \bmod p$$

Then he splits k into k_1 and k_2. and calculates

$$c = E_{k_1}(m), \; r = KH_{k_2}(m) \text{ and } s = x/(r + x_{ap}) \bmod q.$$

The proxy signer sends (c, r, s, K') to Bob.

Proxy unsigncryption: Bob computes

$$y_{ap} = y_a K'^{K'} \bmod p \text{ and } k = (y_{ap}g^r)^{sx_B} \bmod p,$$

then splits k into k_1 and k_2. He recovery the message $m = D_{k_1}(c)$ and accepts m is a valid message of the proxy signer only if the following equation holds:

$$KH_{k_2}(m) = r.$$

Similar to Zheng's original scheme, in Gamage et al.'s proxy signcryption scheme the following equation is holds:

$$k = (y_{ap}g^r)^{sx_B} = \left(y_B^{x_{ap}+r}\right)^s \bmod p.$$

When a third party knows x_{ap}, he can computes k as in the right side of the equation. So this proxy signcryption scheme does not provide forward secrecy property with respect to the proxy signer. Since Alice knows x_{ap}, this scheme does not protect the proxy signer from Alice to forge a proxy signature. Moreover, the proxy signature of this scheme cannot be publicly verified.

1.2 Shin et al.'s DSA-verifiable signcryption Scheme

Signcryption: Alice randomly chooses $x \in Z_q^*$ and carries out the following procedures to signcrypt the message m:

1. $K = y_B^x \bmod p$
2. $K_1 = H(K)$
3. $k = g^x \bmod p$
4. $r = k \bmod q$
5. $h = H(m)$
6. $s = (h + x_A r)/x \bmod q$
7. $e_1 = h/s \bmod q$ and $e_2 = r/s \bmod q$
8. $c = E_{K_1}(m, e_1, e_2)$.

Alice sends (k, e) to Bob.

Unsigncryption: Bob carries out the following procedures to unsigncrypt the message m:

1. $K = k^{x_B} \bmod p$

2. $K_1 = H(K)$
3. $c = E_{K_1}(m, e_1, e_2)$
He accepts m is a valid message of Alice only if the following equation holds:

$$k = g^{e_1} y_A^{e_2} \bmod p.$$

Publicly verification: Bob sets
1. $r = g^{e_1} y_A^{e_2} \bmod p \bmod q$
2. $s = r/e_2 \bmod q$
and sends (m, r, s) to the verifier.
The signature (m, r, s) is a DSA-verifiable signature [9].

2.   The proposed proxy signcryption scheme

Proxy key generation: Alice sends (x_{ap}, K') through a secure channel to the proxy signer. The proxy signer accepts x_{ap} as a valid proxy signature key only if the following equation is holds:

$$g^{x_{ap}} = y_a K'^{K'} \bmod p.$$

Proxy signcryption: The proxy signer randomly chooses $x \in Z_q^*$ and carries out the following procedures to signcrypt the message m:
1. $K = y_B^x \bmod p$
2. $K_1 = H(K)$
3. $k = g^x \bmod p$
4. $r = k \bmod q$
5. $h = H(m)$
6. $s = (h x_P + x_{ap} r)/x \bmod q$
7. $e_1 = h/s \bmod q$ and $e_2 = r/s \bmod q$
8. $c = E_{K_1}(m, e_1, e_2)$.
The proxy signer sends (k, e) to Bob.
Proxy unsigncryption: Bob computes $y_{ap} = y_a K'^{K'} \bmod p$ and carries out the following procedures to signcrypt the message m:
1. $K = k^{x_B} \bmod p$
2. $K_1 = H(K)$
3. $c = E_{K_1}(m, e_1, e_2)$
He accepts m is a valid message of Alice only if the following equation holds:

$$k = y_P^{e_1} y_{ap}^{e_2} \bmod p.$$

Publicly verification: Bob sets
1. $r = y_P^{e_1} y_{ap}^{e_2} \bmod p \bmod q$
2. $s = r/e_2 \bmod q$ and sends (m, r, s, K') to the verifier.

3.   Analysis

The security analysis of our scheme is similar to that of Gamage et al.'s scheme. We only consider the following three problems.

Since the secret key x_P is used in the proxy signcryption phase, only the proxy signer can create a valid proxy signcryption. Bob can assure of the proxy signer's identity. This scheme can protect the proxy signer against the original signer's forgery attack.

Even if x_{ap} is revealed, a person cannot computes $K = k^{x_B} \bmod p$ since he does not know the value of x_B. Therefore, our scheme provides forward secrecy property with respect to the proxy signer.

In a later of dispute, Bob can send (m, r, s, K') to a verifier. The verifier can settle the dispute. Therefore, our scheme is publicly verifiable.

4.   Conclusion

Proxy signcryption scheme is useful for applications that are based on unreliable datagram style network communication model where messages are individually signed and not serially linked via a session key to provide authenticity and integrity. In this paper, we have proposed a new publicly verifiable proxy signcryption scheme. The proposed scheme can overcome some weaknesses of the previous schemes.

References

[1] K. Nyberg and R. A. Rueppel. Message recovery for signature schemes based on the discrete logarithm problem, In: Eurocrypt '94, Springer-Verlag, 1995, pp. 182–193.

[2] Y. Zheng. Digital signcryption or how to achieve cost(signature & encryption)«cost(signature) + cost(encryption), In: CRYPTO '97, Springer-Verlag, 1997, pp. 165–179.

[3] H. Petersen and M. Michels. Cryptanalysis and improvement of signcryption schemes, IEE proc. Comput. Digit. Tech., 1998, 145 (2) pp. 149-151.

[4] F. Bao and R.H. Deng. A signcryption scheme with signature directly verifiable by public key, in PKC'98, Springer-Verlag, 1998, pp. 55 - 59.

[5] H. Jung, D. Lee, J. Lim and K. Chang. Signcryption schemes with forward secrecy. Proceedings of WISA2001, Springer-Verlag, 2001.

[6] J. Shin, K. Lee and K. Shim. New DSA-verifiable signcryption schemes, In ICISC'2002, Springer-Verlag, 2003, pp. 35-47.

[7] G. Gamage, J. Leiwo and Y. Zheng. An efficient scheme for secure message transmission using proxy-signcryption. Technical report 98-01, Monash University, 1998.

[8] M. Mambo, K. Usuda, and E. Okamoto. Proxy signature: Delegation of the power to sign messages. IEICE Trans. Fundamentals, 79 (9): 1338-1353, 1996.

[9] NIST(National Institute for Standard and Techonology). Digital Signature Standard (DSS). FIPS PUB 186, 1994.

SOME NEW PROXY SIGNATURE SCHEMES FROM PAIRINGS

Fangguo Zhang,
Reihaneh Safavi-Naini
School of Information Technology and Computer Science
University of Wollongong, NSW 2522 Australia
{fangguo,rei}@uow.edu.au

Chih-Yin Lin
Computer and Communications Research Laboratories (CCL),
Industrial Technology Research Institute (ITRI), Hsinchu, Taiwan
lincy@itri.org.tw

Abstract   Proxy signatures were first introduced by Mambo, Usuda, and Okamoto. After that, many proxy signature schemes and various types of proxy signature schemes have been proposed. Due to the various applications of the bilinear pairings in cryptography, many ID-based signature schemes have been proposed. In this paper, we propose a general construction of proxy signature with warrant from ID-based signature schemes using bilinear pairings, and give some concrete proxy and proxy blind signature schemes based on existed ID-based signature schemes.

Keywords: proxy signature, ID-based cryptography, proxy blind signatureproxy blind signature, bilinear pairings

1.   Introduction

The concept of proxy signature was first introduced by Mambo, Usuda, and Okamoto in 1996 [9]. The proxy signature schemes allow proxy signers to sign messages on behalf of an original signer. Such signatures have found numerous applications, particularly in distributed computing where delegation of rights is quite common. After Mambo *et al.*'s first scheme was announced, many proxy signature schemes have been proposed. Furthermore, proxy signatures can combine other special signatures to obtain some new types of proxy signatures. Till now, there are various kinds of proxy signature schemes have been proposed, such as threshold proxy signature [16], proxy multi-signature [13], proxy blind signature, ect.

Proxy blind signature is an important type of proxy signature, it plays an important role in the following scenario: In e-cash system, the user makes the bank blindly sign a coin using blind signature schemes. Whenever a user goes through a valid branch to withdraw a coin, he/she needs the branch to make proxy blind signature on behalf of the signee bank.

In the last couple of years, the bilinear pairings have found various applications in cryptography, they can be used to realize some cryptographic primitives that were previously unknown or impractical. More precisely, they are basic tools for construction of ID-based cryptographic schemes (This concept of ID-based public key cryptosystem was first proposed by Shamir [12]), many ID-based cryptographic schemes have been proposed using them [1, 3, 5]. In this paper we address that it is easy to design proxy signature and proxy blind signature from ID-based signature schemes using bilinear pairings, and give some concrete schemes.

The rest of the paper is organized as follows: Section 2 briefly explains some preliminaries. Section 3 gives a description of the general construction of various types of proxy signature from ID-based public key setting using bilinear pairing. In Section 4 and 5, some concrete proxy signature schemes are presented. Section 6 concludes this paper.

2.   Preliminaries

2.1   Bilinear Pairings and BLS Signature Scheme

Let $\mathbb{G}_1$ be a cyclic additive group generated by P, whose order is a prime q, and $\mathbb{G}_2$ be a cyclic multiplicative group of the same order q. A bilinear pairing is a map $e : \mathbb{G}_1 \times \mathbb{G}_1 \to \mathbb{G}_2$ with the following properties:

P1 *Bilinear:* $e(aP, bQ) = e(P, Q)^{ab}$;

P2 *Non-degenerate:* There exists $P, Q \in \mathbb{G}_1$ such that $e(P, Q) \neq 1$;

P3 *Computable:* There is an efficient algorithm to compute $e(P, Q)$ for all $P, Q \in \mathbb{G}_1$.

Throughout this paper, we define the system parameters in all schemes are as follows: Let P be a generator of $\mathbb{G}_1$, the bilinear pairing is given by $e : \mathbb{G}_1 \times \mathbb{G}_1 \to \mathbb{G}_2$. Define two cryptographic hash functions $H_1 : \{0, 1\}^* \to Z_q$ and $H_2 : \{0, 1\}^* \to \mathbb{G}_1$.

Now we are ready to introduce Boneh *et al.*'s pairing-based short signature scheme proposed in [2], we denote BLS scheme.

   1 **Key generation.** Pick random $x \in_R Z_q^*$, and compute $P_{pub} = xP$. The public key is P_{pub}. The secret key is x.

2 **Sign.** A message $M \in \{0,1\}^*$, $P_M = H_2(M) \in \mathbb{G}_1$, $S_M = sP_M$. The signature of M is S_M.

3 **Ver.** Check whether the following equation holds:
$$e(S_M, P) = e(H_2(M), P_{pub}).$$

This scheme is proven to be secure against existential forgery on adaptive chosen-message attacks (in the random oracle model) assuming the CDHP is hard [2].

2.2 Proxy Signature and Proxy Blind Signature

A proxy signature scheme consists of three entities: original signer, proxy signer and verifier. One assumes that each participant has received (via a PKI or a certificate) a public-secret key pair (*Setup*). When the original signer desires to delegate his/her signing ability to the proxy signer, they run a possibly interactive protocol: *Generation of the proxy key*. The proxy signer can use the proxy signature key to sign messages on behalf of the original signer (*Proxy signature generation*). Anyone can verify the validity of such signatures using a proxy *Verification* algorithm.

Depending on whether the original signer can generate the same proxy signatures as the proxy signers do, there are two kinds of proxy signature schemes: (1) Proxy-unprotected ; (2) Proxy-protected.

The *Generation of the proxy key* in proxy signature is a delegation procedure. There are three types of delegation in Mambo *et al.*'s paper: full delegation, partial delegation and delegation by warrant. In [6], S. Kim *et al.* gave a new type of delegation called partial delegation with warrant, which can be considered as the combination of partial delegation and delegation by warrant.

Lee *et al.* [7] defined properties that a strong proxy signature scheme should provide: *Distinguishability, Verifiability, Strong non-forgeability, Strong identifiability, Strong non-deniability and Prevention of misuse.*

Proxy blind signature is considered be the combination of proxy signature and blind signature, so, beside above security requirements of proxy signature, it should satisfy the additional requirements: *Blindness, i.e.*, the signer does not know the content of the message. In general, a proxy blind signature scheme consists of four participants: an original signer, a proxy signer, a user and a verifier, and the following five algorithms, *Setup, Generation of the proxy key, Proxy blind signature generation*, and *Verification*.

2.3 General Process of ID-based Signature Scheme from Pairing

ID-based public key setting involves a KGC (Key Generator Center) and users. The basic operations consists of **Setup** and **Private Key Extraction** (simply **Extract**). When we use bilinear pairings to construct ID-based signature scheme, the general process will be as follows:

- **Setup:** KGC chooses a random number $s \in Z_q^*$ and sets $P_{pub} = sP$. The center publishes system parameters $params = \{\mathbb{G}_1, \mathbb{G}_2, e, q, P, P_{pub}, H_1, H_2\}$, and keeps s as the *master-key*, which is known only by itself.

- **Extract:** A user submits his/her identity information ID to KGC. KGC computes the user's public key as $Q_{ID} = H_2(ID)$, and returns $S_{ID} = sQ_{ID}$ to the user as his/her private key.

- **Signing:** is a probabilistic polynomial time (PPT) algorithm that takes $params$, a private key S_{ID}, and a message m. The algorithm outputs a signature $\sigma(m)$ for m.

- **Verification:** is a PPT algorithm that takes $(params, ID, m, \sigma(m))$ and outputs either accept or reject.

We address that ID-based signature scheme with a trusted KGC can be regarded as a proxy-unprotected proxy signature scheme with multiple proxies. This is obviously: we take the KGC as the original signer, user as the proxy signer. **Extract** can be considered the **Generation of the proxy key**, this is the delegation.

3.   The General Construction

About the delegation function of pairing based cryptosystems, Boneh *et. al* [1] and Chen *et. al* [4] had noted it. If using their delegation to construct proxy signature schemes directly, they are proxy-unprotected proxy signature schemes. To obtain the proxy-protected delegation, we will require the user to make a signature on the same message using BLS short signature. Assume that there are two participants, one called original signer with public key PK_o and secret key s_o, another called proxy signer with public key PK_p and secret key s_p, they have the common system parameters: $\{\mathbb{G}_1, \mathbb{G}_2, e, q, P, H_1, H_2\}$. We describe the delegation in detail as follows:

- The original signer makes a warrant w. There is an explicit description of the delegation relation in the warrant w.

- The original signer computes $So_w = s_o H_2(w)$, and sends w and So_w to proxy signer.

- The proxy signer checks if $e(So_w, P) = e(H_2(w), PK_o)$, if it is right, then computes $S_w = So_w + s_p H_2(w)$.

In fact this is the partial delegation with warrant [6]. So, it is can be regarded as the **Generation of the proxy key** in proxy signature. The proxy secret key is S_w, and the proxy public key is $PK_o + PK_p$. Then the proxy signer can uses any ID-based signature schemes and ID-based blind signature schemes from pairings (takes the ID public key as $H_2(w)$) and secret key as S_w, the public key of KGC as $PK_o + PK_p$) to get proxy signature and proxy blind signature schemes.

Anyone cannot forge an $S_{w'}$ of a warrant w', since the original signer and proxy signer all use BLS short signature scheme to sign warrant, and BLS short signature scheme is proven to be secure. Like the discussion in [8], above delegation need not the secure channel for the delivery of the signed warrant by the original signer, *i.e.*, the original signer can publish w and S_{q_v}. More precisely, any adversary can get the original signer's signature on warrant w. Even this, the adversary cannot get the S_w of the proxy signer, because S_w satisfies $e(S_w, P) = e(H_2(w), PK_o + PK_p)$, and $e(S_{ow}, P) = e(H_2(w), PK_o)$, so, $e(S_w - S_{ow}, P) = e(H_2(w), PK_p)$. This means if the adversary can get the S_w of the proxy signer, then he can forge the BLS signature of the message w with the public key PK_p of proxy signer, this is impossible due to the security of BLS scheme.

4. New Proxy Signature Schemes

In this section, we give a new proxy signature scheme based on Hess' [5] ID-based signature scheme.

[**Setup:**]

The system parameters $params = \{\mathbb{G}_1, \mathbb{G}_2, e, q, P, H_1, H_2\}$, the original signer has public-secret key pair (PK_o, s_o), the proxy signer has public-secret key pair (PK_p, s_p).

[**Generation of the proxy key:**]

After the original signer and the proxy signer finish the process in Section 3, the proxy signer gets a proxy key S_w.

[**Proxy signature generation:**]

For any delegated message m, the proxy signer uses Hess's ID-based signature scheme [5] (takes the signing key as S_w) and obtains a signature (c_p, U_p) as follows:

$$r_p = e(P, P)^{k_p}, k_p \in_R \mathbb{Z}_q^*, \ c_p = H_1(m\|r_p), \ U_p = c_p S_w + k_p P$$

The valid proxy signature will be the tuple $< m, c_p, U_p, w >$.

[**Verification:**]

A verifier can accept this proxy signature if and only if

$$c_p = H_1(m\|e(U_p, P)e(H_2(w), PK_o + PK_p)^{-c_p}).$$

The verification of the signature is justified by the following equations:

$$e(U_P, P)(e(H_2(w), PK_o + PK_p))^{-c_P}$$
$$= e(c_p S_w + k_p P, P)(e(H_2(w), PK_o + PK_p))^{-c_P}$$
$$= e(c_p(S o_w + s_p H_2(w)), P)e(k_p P, P)(e(H_2(w), PK_o + PK_p))^{-c_P}$$
$$= (e(H_2(w), PK_o + PK_p))^{c_P} e(k_p P, P)(e(H_2(w), PK_o + PK_p))^{-c_P}$$
$$= e(P, P)^{k_p} = r_P$$

So, we have:

$$c_P = H_1(m||r_P) = H_1(m||e(U_p, P)e(H_2(w), PK_o + PK_p)^{-c_p}).$$

Due to using the warrant w, it is obvious that our new proxy signature scheme satisfies the requirements stated in Section 2.2. but **strong non-forgeability**. On the other hand, we use Hess's ID-based signature scheme to generate the proxy signature, and it is proven to be secure under the hardness assumption of CDHP and the random oracle model, so the new proxy signature is unforgeable.

Recently, many ID-based signature schemes have been proposed using the bilinear pairings [3, 5, 10, 11]. Like above construction of Hess version, it is easy to construct other proxy signature schemes based on Paterson scheme [10], Cha-Cheon scheme [3] and Sakai-Ohgishi-Kasahara scheme [11].

5.   New Proxy Blind Signature Schemes

The proxy blind signature satisfies the security properties of both the blind signature and the proxy signature, such signature is suitable for many applications where the users' privacy and proxy signature are required. From the ID-based blind signature scheme, we can construct proxy blind signature scheme. The first ID-based blind signature scheme was proposed by Zhang and Kim [14] in Asiacrypt2002. Recently, they gave another ID-based blind signature scheme [15]. Now, we give a new proxy blind signature scheme based on this ID-based blind signature scheme.

[Setup:]

The system parameters $params = \{\mathbb{G}_1, \mathbb{G}_2, e, q, P, H_1, H_2\}$, the original signer has public-secret key pair (PK_o, s_o), the proxy signer has public-secret key pair (PK_p, s_p).

[Generation of the proxy key:]

After the original signer and the proxy signer finish the process in Section 3, the proxy signer gets a proxy key S_w.

[Proxy blind signature generation:]

Suppose that m is the message to be signed.

- The proxy signer randomly chooses a number $r \in_R Z_q^*$, computes $U = rH_2(w)$, and sends U and the warrant w to the user.

- (Blinding) The user randomly chooses $\alpha, \beta \in_R Z_q^*$ as blinding factors. He/She computes $U' = \alpha U + \alpha \beta H_2(w)$ and $h = \alpha^{-1} H_1(m||U') + \beta$, sends h to the signer.

- (Signing) The signer sends back V, where $V = (r + h)S_w$.

- (Unblinding) The user computes $V' = \alpha V$. He/She outputs $\{m, U', V'\}$.

Then (U', V', w) is the proxy blind signature of the message m.
[**Verification:**]
A verifier can accept this proxy blind signature if and only if

$$e(V', P) = e(U' + H_1(m||U')H_2(w), PK_o + PK_p).$$

Like the discussion in [15], our new proxy blind signature scheme can provide the batch verification. This is very important when the number of verifications is considerably large (*e.g.*, when a branch bank issues a large number of electronic coins and the customer wishes to verify the correctness of the coins). Assuming that $(U'_1, V'_1), (U'_2, V'_2), \cdots, (U'_n, V'_n)$ are proxy blind signatures on messages $m_1, m_2, \cdots, m_n$ which issued by the proxy signer with the public key PK_p and the same warrant w form the original signer. The batch verification is then to test if the following equation holds:

$$e(\sum_{i=1}^{n} V'_i, P) = e(\sum_{i=1}^{n} U'_i + (\sum_{i=1}^{n} H_1(m_i, U'_i))H_2(w), PK_o + PK_p).$$

The correctness of the verification is easy to check. A warrant made by the original signer is included in a valid proxy blind signature, so, the proxy blind signature is distinguishable, verifiable, identifiable and non-deniable. The blindness and the non-forgeability of this new proxy blind signature are similar to the discussion of [15].

6. Conclusion

Various type proxy signatures are important in many applications, such as secure e-commerce. Due to the various applications of the bilinear pairings in cryptography, there are many ID-based cryptographic schemes have been proposed. In this paper, we first have shown how we can obtain the proxy-protected delegation using the short signature scheme of Boneh, Lynn and Shacham. Using this delegation, it is easy to design the proxy signature and proxy blind signature from the conventional ID-based signature schemes using bilinear pairings, we have given some concrete schemes based on existed ID-based signature schemes.

References

[1] D. Boneh and M. Franklin, *Identity-based encryption from the Weil pairing*, Crypto 2001, LNCS 2139, pp.213-229, Springer-Verlag, 2001.

[2] D. Boneh, Lynn B. and Shacham H., *Short signatures from the Weil pairing*, In C. Boyd, editor, Asiacrypt 2001, LNCS 2248, pp.514-532, Springer-Verlag, 2001.

[3] J.C. Cha and J.H. Cheon, *An identity-based signature from gap Diffie-Hellman groups*, PKC 2003, LNCS 2139, pp.18-30, Springer-Verlag, 2003.

[4] L. Chen, K. Harrison, A. Moss, D. Soldera and N.P. Smart, *Certification of public keys within an identity based system*, ISC 2002, LNCS 2433, pp. 322-333, Springer-Verlag, 2002.

[5] F. Hess, *Efficient identity based signature schemes based on pairings*, SAC 2002, LNCS 2595, pp.310-324, Springer-Verlag, 2002.

[6] S. Kim, S. Park, and D. Won, *Proxy signatures, revisited*, ICICS 97, LNCS 1334, Springer-Verlag, pp. 223-232, 1997.

[7] B. Lee, H. Kim and K. Kim, *Secure mobile agent using strong non-designated proxy signature*, ACISP2001, LNCS 2119, pp.474-486, Springer Verlag, 2001.

[8] J.Y. Lee, J.H. Cheon and S. Kim, *An analysis of proxy signatures: Is a secure channel necessary?*, CT-RSA 2003, LNCS 2612, pp. 68-79, Springer-Verlag, 2003.

[9] M. Mambo, K. Usuda, and E. Okamoto, *Proxy signature: Delegation of the power to sign messages*, In IEICE Trans. Fundamentals, Vol. E79-A, No. 9, pp. 1338-1353, 1996.

[10] K.G. Paterson, *ID-based signatures from pairings on elliptic curves*, Electron. Lett., Vol.38, No.18, pp.1025-1026, 2002.

[11] R. Sakai, K. Ohgishi, M. Kasahara, *Cryptosystems based on pairing*, SCIS 2000-C20, Jan. 2000, Okinawa, Japan.

[12] A. Shamir, *Identity-based cryptosystems and signature schemes*, Crypto 84, LNCS 196, pp.47-53, Springer-Verlag, 1984.

[13] L. Yi, G. Bai and G. Xiao, *Proxy multi-signature scheme: A new type of proxy signature scheme*, Electronics Letters, Vol. 36, No. 6, 2000, pp.527-528.

[14] F. Zhang and K. Kim, *ID-based blind signature and ring signature from pairings*, Asiacrpt2002, LNCS 2501, pp. 533-547, Springer-Verlag, 2002.

[15] F. Zhang and K. Kim, *Efficent ID-based blind signature and proxy signature from pairings*, ACISP 2003, LNCS 2727, pp. 312-323, Springer-Verlag, 2003.

[16] K. Zhang, *Threshold proxy signature schemes*. 1997 Information Security Workshop, Japan, Sep., 1997, pp.191-197.

CONSTRUCTION OF DIGITAL SIGNATURE SCHEMES BASED ON DISCRETE LOGARITHM PROBLEM*

Wei-Zhang Du
School of Computer Science & Technology, Yantai University
Yantai 264005, China
duweizhang20@263.net

Kefei Chen
Department of Computer Science & Engineering, Shanghai Jiaotong University,
Shanghai 200030, China
kfchen@mail.sjtu.edu.cn

Abstract   Both the Schnorr digital signature scheme and the Okamoto digital scheme are important digital signature schemes, they are constructed based on the discrete logarithm problems. On the bases of these two digital signature schemes, two new digital signature schemes are constructed in this paper. For the second digital signature scheme, its blind digital signature scheme is given also.

Keywords: digital signature, discrete logarithms problem

1.   Introduction

Both the Schnorr digital signature scheme [1] and the Okamoto digital scheme [2] are important digital signature schemes, they are constructed based on the discrete logarithm problems. The Schnorr digital signature scheme and the Okamoto digital signature scheme and their blind digital signature scheme have found wide use. On the bases of these two digital signature schemes, based on the discrete logarithm problem, two new digital signature schemes are constructed in this paper. For the second digital signature scheme, its blind digital signature scheme is given also.

*This work was supported by NSFC under grant #90104005 and #60273049

2.   Constructions of schemes

2.1   Protocol One

Suppose p, q are prime, $q \mid p - 1$, q is about 140 bits, p is at least 512 bits, g_1, g_2 are random numbers with the same length of q. Let $H(x)$ be a collision-resistant one-way hash function that maps $\{0, 1\}^*$ to Z_q. x_1, x_2 are the signer's secret keys, both of them are random numbers which are less than q. $y = g_1^{x_1} g_2^{x_2} \bmod p$ is the signer's public key. The signing process is as follows:

(1) The signer Alice selects two random numbers k_1, k_2 which are less than q;

(2) Alice computers

$$r = \left(g_1^{k_1} g_2^{k_2} \bmod p\right) \bmod q$$
$$s_1 = \left(k_1^{-1} \left(H\left(m\right) + x_1 r\right)\right) \bmod q$$
$$s_2 = \left(k_2^{-1} \left(H\left(m\right) + x_2 r\right)\right) \bmod q$$

(r, s_1, s_2) are signatures of the signer Alice. She sends them to the recipient Bob.

(3) Bob verify the signature by computing:

$$w_1 = s_1^{-1} \bmod q, \quad w_2 = s_2^{-1} \bmod q,$$
$$u_1^1 = (H\left(m\right) \times w_1) \bmod q, \quad u_1^2 = (H\left(m\right) \times w_2) \bmod q,$$
$$u_2^1 = (r w_1) \bmod q, \quad u_2^2 = (r w_2) \bmod q,$$
$$v = \left(\left(g_1^{u_1^1} \cdot g_2^{u_1^2} \cdot y_1^{u_2^1} \cdot y_2^{u_2^2}\right) \bmod p\right) \bmod q,$$

where $y_1 = g_1^{x_1} \bmod p$, $y_2 = g_2^{x_2} \bmod p$. If $v = r$, then the signature is valid.

(4) Verification of signature:

$$\left(\left(g_1^{u_1^1} \cdot g_2^{u_1^2} \cdot y_1^{u_2^1} \cdot y_2^{u_2^2}\right) \bmod p\right) \bmod q$$
$$= \left(\left(g_1^{H(m) \cdot k_1 (H(m) + x_1 r)^{-1}} \cdot g_1^{x_1 r k_1 (H(m) + x_1 r)^{-1}} \cdot g_2^{H(m) \cdot k_2 \cdot (H(m) + x_2 r)^{-1}}\right.\right.$$
$$\left.\left. \cdot g_2^{x_2 \cdot r \cdot k_2 \cdot (H(m) + x_2 r)^{-1}}\right) \bmod p\right) \bmod q$$
$$= \left(\left(g_1^{k_1 (H(m) + x_1 r)^{-1} \cdot (H(m) + x_1 r)} \cdot\right.\right.$$
$$\left.\left. g_2^{k_2 (H(m) + x_2 r)^{-1} \cdot (H(m) + x_2 r)}\right) \bmod p\right) \bmod q$$
$$= \left(\left(g_1^{k_1} \cdot g_2^{k_2}\right) \bmod p\right) \bmod q$$

2.2　Protocol Two

Suppose p, q are prime, $q \mid p - 1$, q is about 140 bits, p is at least 512 bits, g_1, g_2, g_3 are random numbers with the same length of q. Let $H(x)$ be a collision-resistant one-way hash function that maps $\{0, 1\}^*$ to Z_q. s_1, s_2, s_3 are the signer's secret keys, all of them are random numbers which are less than q. $v = g_1^{-s_1} g_2^{-s_2} g_3^{-s_3} \bmod p$ is the signer's public key. Given message M, the signing process is as follows:

(1) The signer Alice selects three random numbers r_1, r_2, r_3 which are less than q;

(2) Alice computers

$$
\begin{aligned}
e &= H\left(g_1^{r_1} g_2^{r_2} g_3^{r_3} \bmod p, M\right), \\
y_1 &= (r_1 + es_1) \bmod q, \\
y_2 &= (r_2 + es_2) \bmod q, \\
y_3 &= (r_3 + es_3) \bmod q,
\end{aligned}
$$

(e, y_1, y_2, y_3) are signature of the signer Alice to the message M. She sends them to the recipient Bob.

(3) Bob verify the signature :

$Ver\left(M, e, y_1, y_2, y_3\right)$ is true if and only if

$$
H\left(g_1^{y_1} g_2^{y_2} g_3^{y_3} v^e \bmod p, M\right) = e
$$

Verification of signature:

$$
\begin{aligned}
&\left(g_1^{y_1} g_2^{y_2} g_3^{y_3} \cdot v^e\right) \bmod p \\
= &\left(g_1^{r_1+es_1} \cdot g_2^{r_2+es_2} \cdot g_3^{r_3+es_3} \cdot \left(g_1^{-s_1} g_2^{-s_2} g_3^{-s_3}\right)^e\right) \bmod p \\
= &\left(g_1^{r_1} g_2^{r_2} g_3^{r_3}\right) \bmod p
\end{aligned}
$$

2.3　Blind Protocol of the Protocol Two

For the second protocol, we are able to construct its blind digital signature scheme. Suppose p, q are prime, $q \mid p - 1$, q is about 140 bits, p is at least 512 bits, g_1, g_2, g_3 are random numbers with the same length of q. Let $H(x)$ be a collision-resistant one-way hash function that maps $\{0, 1\}^*$ to Z_q. s_1, s_2, s_3 are the signer's secret keys, all of them are random numbers which are less than q. $v = g_1^{-s_1} g_2^{-s_2} g_3^{-s_3} \bmod p$ is the signer's public key. Given message M, the signing process is as follows:

Let signer be Σ, verifier be V, then the protocol can be shown as following Fig. 1.

To show that the protocol is blind it suffices to show that for every possible view of the signer and for every possible signature there exists exactly one suitable quadruple of blinding factors $(\alpha_1, \alpha_2, \alpha_3, \beta)$. Given a view consisting

V	Σ
(M, g_1, g_2, g_3, v)	$(g_1, g_2, g_3, s_1, s_2, s_3)$

$$r_1, r_2, r_3 \in_R Z_q^*$$
$$t' := g_1^{r_1} g_2^{r_2} g_3^{r_3} \bmod p$$

$\xleftarrow{\quad t' \quad}$

$$\alpha_1, \alpha_2, \alpha_3, \beta \in_R Z_q$$
$$t := t' g_1^{\alpha_1} g_2^{\alpha_2} g_3^{\alpha_3} \cdot v^\beta \bmod p$$
$$e := H(t, M)$$
$$e' := e - \beta \,(\bmod q)$$

$\xrightarrow{\quad e' \quad}$

$$y_1' = r_1 + e' s_1 \,(\bmod q)$$
$$y_2' = r_2 + e' s_2 \,(\bmod q)$$
$$y_3' = r_3 + e' s_3 \,(\bmod q)$$

$\xleftarrow{\quad y_1', y_2', y_3' \quad}$

$$y_1 := y_1' + \alpha_1 \,(\bmod q)$$
$$y_2 := y_2' + \alpha_2 \,(\bmod q)$$
$$y_3 := y_3' + \alpha_3 \,(\bmod q)$$

$$e \stackrel{?}{=} H(g_1^{y_1} g_2^{y_2} g_3^{y_3} v^e \bmod p, M)$$

If true, getting a blind signature of M :

$$(e, y_1, y_2, y_3)$$

Fig 1. Blind Signature Protocol Based on the Protocol Two

of $r_1, r_2, r_3, t', e', y_1', y_2', y_3'$ and a signature (e, y_1, y_2, y_3) on a message M, the only possibility is

$$
\begin{aligned}
\alpha_1 : &= y_1 - y_1' \,(\bmod q) \\
\alpha_2 : &= y_2 - y_2' \,(\bmod q) \\
\alpha_3 : &= y_3 - y_3' \,(\bmod q) \\
\beta : &= e - e' \,(\bmod q)
\end{aligned}
$$

With these blinding factors, the verifier V would have computed

$$
\begin{aligned}
t : &= t' g_1^{\alpha_1} g_2^{\alpha_2} g_3^{\alpha_3} \cdot v^\beta \\
e : &= H(t, M)
\end{aligned}
$$

It remains to show that $t = g_1^{y_1} g_2^{y_2} g_3^{y_3} v^e$:

$$
\begin{aligned}
t &= t' g_1^{\alpha_1} g_2^{\alpha_2} g_3^{\alpha_3} \cdot v^\beta \\
&= g_1^{r_1} g_2^{r_2} g_3^{r_3} g_1^{\alpha_1} g_2^{\alpha_2} g_3^{\alpha_3} \cdot v^\beta \\
&= g_1^{r_1} g_2^{r_2} g_3^{r_3} g_1^{y_1 - y_1'} g_2^{y_2 - y_2'} g_3^{y_3 - y_3'} \cdot v^{e-e'} \\
&= g_1^{y_1} g_2^{y_2} g_3^{y_3} \cdot v^e \cdot g_1^{r_1 - y_1'} g_2^{r_2 - y_2'} g_3^{r_3 - y_3'} \cdot v^{-e'} \\
&= g_1^{y_1} g_2^{y_2} g_3^{y_3} \cdot v^e \cdot g_1^{-e' s_1} g_2^{-e' s_2} g_3^{-e' s_3} \cdot \left(g_1^{-s_1} g_2^{-s_2} g_3^{-s_3} \right)^{-e'} \\
&= g_1^{y_1} g_2^{y_2} g_3^{y_3} \cdot v^e
\end{aligned}
$$

Thus, $e = H\left(g_1^{y_1} g_2^{y_2} g_3^{y_3} \cdot v^e \bmod p, M \right)$ and the blinding factors $(\alpha_1, \alpha_2, \alpha_3, \beta)$ would in fact have resulted in the valid signature.

3. Conclusion

On the bases of Schnorr signature scheme and Okamoto signature scheme, we have suggested two digital signature schemes based on the discrete logarithm problems. For the second scheme, we have given its blind signature scheme also. We believe that these schemes will be widely used in the real world.

References

[1] [1] C. P. Schnorr, Efficient signature generation by smart cards, Journal of Cryptology, 1991, Vol.4, No.3, pp.161-174.

[2] [2] T. Okamoto, Provably secure and practical identification schemes and corresponding signatures schemes, Advances in Cryptology-Crypto'92, Lecture notes in computer science, Vol.740, Springer-Verlag, 1993, pp.31-53.

HOW TO CONSTRUCT DLP-BASED BLIND SIGNATURES AND THEIR APPLICATION IN E-CASH SYSTEMS

Weidong Qiu

Dept. of Communication Systems
University of Hagen
58084 Hagen, Germany
weidong.qiu@fernuni-hagen.de

Abstract   Since the concept of blind signature was first introduced by Chaum, there are many applications of blind signatures. Especially, blind signatures have been applied widely in anonymous E-Cash systems. The most used blind signatures in E-Cash systems are based on discrete logarithm problem (DLP in short).

This paper investigates the method to construct DLP-based blind signatures and tries to generalize how can these blind signatures be utilized to build double-spending resistant anonymous E-Cash systems.

Keywords: blind signature, discrete logarithm problem, electronic cash system

1.   Introduction

Blind signature was first introduced by Chaum in 1982 [8]. Blind signatures allow a recipient to have a signer signed a message m without revealing any information about the message to the signer. This feature can be utilized to provide anonymity for many applications, i.g. electronic cash(E-Cash) systems and electronic voting systems etc. Most of the presented E-Cash systems rely on blind signature for providing anonymity.

The first blind signature was based on RSA system (factoring problem) [8]. It is well known that electronic coins are bit strings which are vulnerable to be copied and spent more than once (double-spending problem). To avoid double-spending problem, Chaum's system was on-line. The bank prevents double-spending by on-line checking whether the coin has been spent or not. Obviously, on-line searching a database(the database may be very large to an unacceptable extent) to make the system double-spending resistant is very impractical. Chaum proposed an off-line E-Cash system still using RSA blind signature based on factoring problem in [9]. The system is able of revealing the identity of the owner of the double spent coins "after the fact". But the system is still quite inefficient for using cut-and-choose technique. Later, more efficient DLP-based blind signatures were presented and consequently, more efficient E-Cash systems based on these blind signatures were designed [11, 10, 6, 7, 4, 18, 5, 3, 19, 9]. Among these systems, the most significant concept is the restrictive blind signature proposed by Brand [3]. This concept has been playing a very important role in the area of E-Cash for the last decade. Numerous systems are derived from or based on the restrictive blind siganture.

In this paper, we illustrate with some examples to show how to construct DLP-based blind signatures and generalize the use of blind signatures in anonymity revocable E-Cash systems. We conclude that knowing the process of constructing DLP-based blind signatures and the principle of their use in E-Cash systems is valuable for designing new efficient E-Cash systems or new blind signatures and group signatures.

The rest of this paper is structured as follows. Section 2 illustrates how to construct DLP-based blind signatures with some concrete examples. Section 3 lists the blinding equations used in some DLP-based blind siguntures. Section 4 simply describes some models for applying blind signatures in the area of E-Cash. We conclude this paper in the last section.

2.   How to construct DLP-based blind signatures

The first DLP-based blind signatures applied into E-cash systems can be found in [9, 3]. In the context of a blind signature, blindness is a very important aspect. We give the definition of blindness of a bind signature [5] as follows.

Def.1. *If the signer's view of a signature execution and the signature results on message, are statistically independent, the signature scheme is called blind.*

We note that blindness of a blind signature indicates that the signature-message pairs are unlinkable. In other words, even knowing N valid signature-message pairs, no one except the signer can consturct the $(N + 1)$th valid signature-message pair. Therefore, the E-Cash systems built on blind signatures are also unlinkable as the bank notes or coins we use in our real life (we can not decide whether two diffenent paper currencies or coins come from the same customer or not).

Normally, in an DLP-based signature or authentication scheme, to prove the knowledge of a secret but not revealing any information about it, a signer or a prover has to compute the ordinate of a point of a line. The intercept and the abscissa of the line are chosen at random by a signature acquirer or/and by the signer. The slope of the line is the secret (a discrete logarithm w.r.t a base g) which is known only to the prover or signer. With the public key of the signer, any recipients can verify the signature to make sure that the signer knows the secret. Under DLP assumption, no one can derive information about the secret from the signatures no matter how many times the signature prototocl or authentication protocol has been executed. This is a zero knowledge proof precess. To construct blind signatures based on DLP, the signer's view of the protocol must be randomized to obtain the blindness of a blind signature.

Proposition 1. *The DLP-based blind signatures can be constructed by randomizing any coefficients of the line (except the slope of the line, i.e. the secret) which can be viewed by the signer during the normal knowledge proof process. Afterwards, the rest of coefficients of the line can be deduced under some equalities satisfying the normal signature process. Which coefficent should be selected to be randomized can be decided according to feasibility or efficiency.*

Almost all this kind of blind signatures [9, 3, 5, 19, 7, 22] can be built up in this way. We illustrate the process of constructing the sort of blind signatures with two examples given separately in [3] and [5].

Stefan Brand's restrictive blind signature Restrictive blind signature was first introduced by Brand [3] and is very suitable for designing double-spending resistant or fair off-line E-Cash systems. To some extent, it is a basic model for constructing practical E-Cash systems and plays important role in the area of E-Cash. We investigate the process of constructing such a blind signature. We follow the denotations defined in Brand's scheme.

system parameters
User: $u_1 \in_R Z_q$, $I = g^{u_1}$ is related to and stored together with the user's identifying information at the bank.
Bank: computes $z = (Ig_2)^x$ and sends it to the user
Normal signing process
The normal process of Brand's signature scheme is shown in figure 1.
Blinding process
Now we explain how to construct the blind signature on $(Ig_2)^s$ instead of Ig_2 as in the normal signing process.

- First, according to the definition of the blindness in Def.1, there should be a new signature tuple (r', c', z', a', b') on the blinded message $(Ig_2)^s$. The two tuples, (r, c, z, a, b) and (r', c', z', a', b') should be completely independent and satisfy the verification equations $g^r = h^c a$, $(Ig_2)^r = z^c b$, and $g^{r'} = h^{c'} a'$, $(Ig_2)^{r'} = z'^{c'} b'$ separately. (r, c, z, a, b) is the

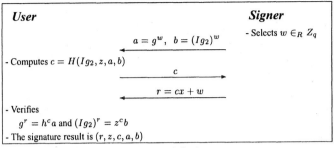

Figure 1.   Normal signing process

signer's view during the signing process and (r', c', z', a', b') is the final blind signature result which will be not known to the signer.

- To blind the signing process, r can be first randomized according to the proposition 1 by setting $r' = ru + v$ as in Brand's scheme [3] or $r' = (r + u)v$ as in Chaum's scheme [9], where integers u, v are chosen at random. Obviously, r and r' are independent because of random chosen u, v.

- The rest of the tupe (r', c', z', a', b') can be dedueced now as follows. (r, c, z, a, b) is the tuple seen by the signer satisfying $g^r = h^c a, (Ig_2)^r = z^c b$ and $r = cx + w$ where x is the secret known only to the signer. From the fact that $g^r = h^c a$, to satisfy the equality $g^{r'} = g^{cux} g^{wu+v} = h^{cu} a^u g^v = h^{c'} a'$, the following equalities must hold: $c' = cu, a' = a^u g^v$.

  Similarly, z', b' can be deduced in the same way: to satisfy the equality $(Ig_2)^{sr'} = (Ig_2)^{scux} (Ig_2)^{swu+v} = z^{scu} (Ig_2)^{swu+sv} = z'^{c'} b', z' = z^s$ (from above, we already know that $c' = cu$), and $b' = ((Ig_2)^w)^{su} ((Ig_2)^s)^v = b^{su} A^v$ must hold. The blind signature then has been constructed.

Another example J.Camenisch presented two DLP-based blind signatures in [5]. One of them is based on a modification of DSA. Another one is derived from Nybeg-Rueppel scheme [16]. We describe the blinding process of the later blind signature as another example to show how can we obtain this sort of blind signatures.

system parameters The system parameters are as follows. A prime p and $p - 1$ has a large prime factor q. An element $g \in Z_p^*$ of order q. The system's private key is $x \in Z_q$, the corresponding public key is $y = g^x \bmod p$. To sign a message $m \in Z_p$, a signer selects $k \in Z_q$ at random and computes c, s as follows.

- $c = mg^k \bmod p$

- $s = xc + k \bmod p$

The pair (c, s) is the signature on the message m. Any recipients can check the equality $m = g^{-s}y^c c \bmod p$ to verify the signature.

Blinding process To obtain the blind signature (c', s') satisfying the equality $m = g^{-s'}y^{c'} c' \bmod p$, where (c', s') and (c, s) should be independent, s can be randomized with the line $s' = s\alpha + \beta$, $\alpha, \beta \in Z_q^*$ are randomly selected integers. s will be computed by the signer with form of $s = \eta x + k$. k is chosen by the signer at random. Substituting $s' = s\alpha + \beta$ in the $g^{-s'}y^{c'} c' \bmod p$, to satisfy the equality $g^{-s'}y^{c'} c' = g^{-s\alpha-\beta+xc'} c' = g^{-\eta\alpha x - k\alpha - \beta + xc'} c' = m \bmod p$, c' must equal to $mg^{k\alpha+\beta}$ and $\eta = c'\alpha^{-1}$. Then, we have obtained the blind signature as described in [5].

3. Generalize some DLP-based blinding processes

In this section, we list some blinding processes from different blind signatures in the following table.

Table 1. Some examples of blinding processes in the different DLP-based blind signatures

Scheme	Blinding factors	Blinding equation
Schnorr-T.Okamoto [19]	u	$y^* = y + u$, $e = e^* + d$
T.Okamoto [19]	u_1 , u_2	$y_1^* = y_1 + u_1$, $y_2^* = y_2 + u_2$, $e = e^* + d$
D.Chaum [9]	u , v	$r = (r_0 + v)u$
S.Brands [3]	u , v	$r' = ru + v$
J.Camenisch [1] [5]	α , β	$R = \tilde{R}^\alpha g^\beta$ (actually blind $\tilde{k}$ by $\alpha\tilde{k} + \beta$)
J.Camenisch [2] [5]	α , β	$s = \tilde{s}\beta + \alpha$
P.Horster [12]	a , b	$a\tilde{k} + b$
H.G.Zhang[22]	a , b	$S' = S + aP_{pub}$, $c = c' + b$

From the table, we can clearly see that, as we pointed out in proposition 1, the general way to construct a DLP-based blind signature is to blind one of the random integers which can be seen by the signer during the signature process.

Blind signatures can protect the privacy of users. But on the other hand the feature of unconditional anonymity can also be misused by criminals for blackmailing or money laundering. In recent years, as the progress on E-Cash systems proceeds, new types of DLP-based blind signatures, for instance fair blind signature [4, 18], partially blind signature [2], partially restrictive blind signatures [14] etc., have been presented and discussed widely in the area of E-Cash systems. Revocable anonymity as well as exact payment can be achieved with these new types of blind signatures. Most of them, especially DLP-based systems, can be built with the same measure described above or with slight variations.

4.   The application of blind signatures in E-Cash

In the progress of E-Cash systems, there are four basic models for the application of blind signatures.

Model I : Normal blind signature Under this simplest model, a bank signs blindly on the coins and checks the coins on-line to prevent multi-spending.

Model II : Restrictive blind signature Brand's restrictive blind signature plays a significant role in developing off-line E-Cash systems. In previous off-line systems which providing anonymity, double-spending resistance, most of them utilize the property of the restrictive blind signature system more or less [3, 18, 4, 7, 6, 10]. Under this model, the principle of utilizing the restrictive blind signature to build E-Cash systems is that a user's identity will be embedded into "inside" construction of a restrictive blind signature which will not be known to the bank. When spending the coin at a merchant, the user proves to the merchant her knowledge on the "inside" construction using the zero knowledge proof. When double spending the coin, two points of a line in the zero knowledge proof will be exposed, and the coefficients of the line can then be computed and used to reveal the "inside" construction of message. Consequently, knowing the "inside" construction results in revealing the the identity information of the user.This kind of system is still anonymous for the bank blindly signing the "outside" construction of the message. This concept can be further extended to built up group signature and group signature based E-Cash systems [20, 15, 21, 17].

Model III : Fair blind signature Although fair blind signature itself is a new concept introduced in [18, 4], it can be easily obtained from restrictive blind signature or other blind signatures and can be further used to construct fair off-line E-Cash systems. Fair E-Cash systems can offer a compromise between the need of the privacy protection of users and effectively preventing the misuse by criminals. The trick in fair blind signatures is that a third party(may be more than one), or called trustee, is involved in the systems. In early systems [4, 7], trustees view all or parts of the blinding process so that the trustees can revoke the anonymity provided by the blind signature. But the trustees has to be involved in each withdrawal or opening account protocol. The efficiency is low. In later systems [6, 10], the trustees have a public-private key pair so that there is no need for the trustees to be on-line or invovled in any protocols except tracing protocol.

Model IV : Partially fair blind signature More recently, a signature called restrictive partially blind signature is proposed by Maitland [14]. Partially blind signatures were introduced by Masayuki Abe [2]. A partially blind signature scheme allows a signer to produce a blind signature on a message while some common agreed information(i.e. expiry date, denominational information) remains visible despite the blinding process. There is no need to use different signing keys for different denominations. We point out here that it is possible to construct anonymity revocable off-line E-Cash which can make exact payment while keep double-spending resistant with restrictive blind partially signature [14, 1, 13]. Exact payment can also imply that there is no need to design divisible E-Cash systems in which complicated cryptographic technologies have to be used resulting in low inefficiency and impractical systems for small amount fund transfer.

For the space limitation, we didn't illustrate the application of different blind signatures in E-Cash systems with concrete examples. How to construct fair or partially blind signatures will be discussed in the full version of this paper.

5. Conclusion

In this paper, we generaized the process of constucting a DLP-based blind signature. Knowing this process, we can convert most of DLP-based digital signature into blind version. Meanwhile, we roughtly described how to utilize different blind signatures to design different E-Cash systems.

References

[1] M. Abe and E. Fujisaki. How to date blind signatures. In *Advances in Cryptology - ASIACRYPT '96, International Conference on the Theory and Applications of Crypotology and Information Security 1996, Proceedings*, LNCS 1163, Springer-Verlag, 1996.

[2] M. Abe and T. Okamoto. Provably secure partially blind signatures. In *Advances in Cryptology - CRYPTO 2000 – 20th Annual International Cryptology Conference, Proceedings*, LNCS 1880, Springer-Verlag, 2000.

[3] S. Brands. Untraceable off-line cash in wallet with observers. In *Advances in Cryptology - CRYPTO '93*, LNCS 773, Springer-Verlag, 1993.

[4] E. Brickell, P. Gemmell, and D. Kravitz. Trustee-based tracing extensions to anonymous cash and the making of anonymous change. In *Proceedings of the 6th Annual ACM-SIAM Symposium on Discrete Algorithms*, ACM, Jan 1995.

[5] J. Camenisch. Blind signatures based on the discrete logarithm problem. In *Advances in Cryptology - EUROCRYPT '94*, LNCS 950, Springer-Verlag, 1994.

[6] J. Camenisch, U.M. Maurer, and M. Stadler. Digital payment systems with passive anonymity-revoking trustees. In *ESORICS'96*, 1996.

[7] J. Camenisch, J. Piveteau, and M. Stadler. An efficient fair payment system. In *Proc.ACM Conference on Computer and Communications Security*, 1996.

[8] D.Chaum. Blind signatures for untraceable payments. In David Chaum, Ronald L.Rivest, and Alan T.Sherman, editors, *Advance in Cryptology: Crypto 82*, Plenum Press, 1983.

[9] D.Chaum and T.P. Pedersen. Wallet databases with observers. In Ernest F.Brickell, editor, *Advance in Cryptology: Proceedings of Crypto 92*, pages 1–14. Springer-Verlag, 1993.

[10] Y. Frankel, Y. Tsiounis, and M. Yung. Indirect discourse proofs: achieving efficient fair off-line e-cash. In *Advances in Cryptology - ASIACRYPT'96*, LNCS 1163, Springer-Verlag, 1996.

[11] (Matthieu Gaud and Jacques Traoré. On the anonimity of fair off-line e-cash systems. In *Financial Cryptography – Seventh International Conference, FC'2003 Proceedings*, LNCS 2742, Springer-Verlag, 2003.

[12] P. Horster, M. Michels, and H. Petersen. Efficient blind signature schemes based on the discrete logarithm problem. Technical Report TR-94-6, University of Technology Chemnitz-Zwickau, 1994.

[13] M.Abe and J.Camenisch. Partially blind signature schemes. In *Proceedings of the 1997 SCIS, SCIS'97-33D*, 1997.

[14] G. Maitland and C. Boyd. A provably secure restrictive partially blind signature scheme. In *Public Key Cryptography, Fourth International Workshop on Practice and Theory in Public Key Cryptography, PKC '02*, LNCS 2274, Springer-Verlag, 2002.

[15] Greg Maitland and Colin Boyd. Fair electronic cash based on a group signature scheme. In *ICICS'01*, LNCS 2229, Springer-Verlag, 2001.

[16] K. Nyberg and R. A. Rueppel. A new signature scheme based on the dsa giving message recovery. In *Conference on Computer and Communications Security – CCS'93*, ACM Press, 1993.

[17] Jacques Traoré Sébastien Canard. On fair e-cash systems based on group signature schemes. In *ACISP'03*, LNCS 2727, Springer-Verlag, 2003.

[18] M. Stadler, J.M. Piveteau, and J.Camenisch. Fair-blind signatures. In *Advances in Cryptology - EUROCRYPT '95*, LNCS 921, Springer-Verlag, 1995.

[19] T.Okamoto. Provable secure and practical identification schemes and corresponding signature signature schemes. In Ernest F.Brickell, editor, *Advance in Cryptology: Proceedings of Crypto 92*, LNCS 740, Springer-Verlag, 1993.

[20] J. Traoré. Group signature and their relevance to privacy-protecting offline electronic cash systems. In *ACISP99*, LNCS 1587, Springer-Verlag, 1999.

[21] W.Qiu, K.Chen, and D.Gu. A new off-line privacy protecting e-cash system with revokable anonymity. In *ISC'02*, LNCS 2433, Springer-Verlag, 2002.

[22] Fangguo Zhang and Kwangjo Kim. Id-based blind signature and ring signature from pairings. In *Advances in Cryptology - ASIACRYPT'02*, LNCS 2501, Springer-Verlag, 2002.

A GROUP OF THRESHOLD GROUP-SIGNATURE SCHEMES WITH PRIVILEGE SUBSETS*

Chen Weidong

Key Lab. of Inform. Security, Graduate School, Chinese Academy of Science, Beijing 100039
Institute of Electronics, Chinese Academy of Sciences, Beijing 100080, China

Feng Dengguo

Key Lab. of Inform. Security, Inst. of Software, Chinese Academy of Sciences, Beijing, China
fdg@is.iscas.ac.cn

Abstract   Reference[9] proposed a threshold group-signature scheme in order to solve the problem so called "threshold group-signature scheme with privilege subsets" suggested by Feng Dengguo. We firstly show there exist some insufficiencies and potential hazard in the scheme mentioned above. Secondly, Using the idea of constructing group-signature schemes by individual signature schemes, we put forward a group of the ones with four variants of ElGamal type, having many attractive properties such as shorter length of signature, message recovery, authentication and so on. Finally, the security of our schemes is proved in the standard model.

Keywords: threshold group-signature scheme; secret sharing scheme; ElGamal cryptosystem; message recovery; provable security

1.   Introduction

The central task of cryptography is privacy and authentication insured and signature is one of the most important mechanism providing authentication. The common signature schemes,such as RSA[1], ElGamal[2], are realized by one signer using his private key, and called "individual signature". However, in many applications, the responsibility of signing is requested to be shared. So, it is natural to introduce the concepts of threshold group-signature: Verifying the validness of signature needs the group public key and anonymity is also asked for. In 1991, Desmedt and Frankel firstly proposed a threshold group-signature

*Supported by the National Gtand for Fundamental Research 973 Program of China under Grant No. G1999035802; the National Natural Science Foundation of China under Grant No. 60253027

scheme based on RSA[3], after that many similar schemes are put forward, such as [4, 5, 6, 7].

One potential problem of schemes mentioned above is that the responsibility of each member is same, but the practical cases are not always as this. In 2000, Feng Dengguo suggested a problem of threshold group-signature with privilege subsets [8]: group G made of n members, has m disjoint subsets, each consisting of $n_i (i = 1, 2, ..., m)$ members. Only when at least t_i members in each subset accept and total number of participants of G t at least, can the group signature be generated. In addition, anonymity and tracing the respective signers in case of authorized are required too.

[8]gives a threshold group-signature scheme satisfying the need above and by far there are not other similar schemes proposed. Of course, some threshold key management protocols borrowing the idea mentioned above, but most are not secure.

We firstly show that the scheme proposed by [8] has many disadvantages. Furthermore, there are some potential hazards on security. Using the idea of constructing group signature schemes by individual signature ones[4], we put forward a group of schemes of ElGamal type, having many attractive properties, such as shorter length of signature, not requiring the assistance of trusted party, simpler realization, authentication (i.e., SC may verifies the pieces submitted by respective members.) and so on. Consisting of four variants, two of them have the property of message recovery, i.e.the message can be recovered from the respective signature, which convenience the application greatly, for example, economical use of bandwidth. We also give the security proof of our schemes.

2.   Analysis on threshold scheme [8]

[8] proposed a threshold group-signature scheme with privilege subsets based on DLP(Discrete Logarithm Problem), called (t_j, t, n)−threshold scheme.

The basic frame of the scheme is as follows.

1) Initialization IDC(trusted identity distribution center) and every member generate identities together and the former also produces the needed parameters. After 12 mutual steps, user i(=1,2,...,n) gets

$$\{id_i, \ two \ parameters \ for \ signing\},$$

and DC(DigitalSignature Combination Center) gains polynomial

$$F_j(x) = \prod_{i \in G_j} (x - id_i) \mod q$$

2) Signing Each member and DC carry out mutual subprotocol of 12 steps. Then DC determines the validity of the identity of each member using respective polynomial $F_j(x)$. Satisfying the need of "privilege threshold", DC broadcasts the group signature parameter $\{j, g_j(x), E_j\}$, where $g_j(x)|F_j(x)$, i.e., is discriminant of identity of signers. Finally, DC collects all individual signatures and produces group signature $\{ID, S, g(x), R_j, E_j | j = 1, 2, ..., m\}$, where,

$$g(x) = \prod_{j=1}^{m} g_j(x) \mod q$$

Obviously, $g(x)$ suggests the identities of participants of G.

3) Verifying omitted.

Analysis as follows. Firstly, protocols involved in is very complicated, especially the longer length of signature. In fact, the length of the signature is not smaller than $(m + 1)[\log_2 p] + (m + t + 1)[\log_2 q]$, where, $q|(p - 1)$, q, p are both secure primes and m is the number of privilege subsets.

Furthermore, the scheme above has potential hazard on security. [8] shows, according to the need of anonymity, the identities of member keep secret, i.e., it should be very difficult to factorize the polynomial $g(x)$. But, in fact there is efficient algorithm[11], such as Berleknmp, etc., for the factorization of polynomial in some finite field, even if the characteristic of field is larger.

3. $(t_1, n_1; ...; t_m, n_m; t, n)$-threshold group-signature

3.1   Basic idea

For the sake of simplicity, we assume group G has only one privilege subset G_1. Our construction of schemes adopts the idea of [4]: Based on mature individual signature schemes, for instance, schemes of ElGamal type, construct threshold group-signature scheme using Secret Sharing Scheme(SSS, such as Shamir's scheme[12]). Compared with [4], we unite the "privilege threshold conditions" with SSS, i.e., Adopting "double secret sharing(double SSS)" idea, the scheme with four variants based on DLP is constructed. Similarly, it may not need the assistance of KAC.

3.2   Initiation

The agencies involved in as follows.

KAC: Trusted key authentication center, responsible for issuing key.

SC: Signature clerk.

G: party consisting of n members.

G_1: Privilege subset.

KAC Operates as follows.

1) Selects two "secure" primes p, q, where, $q|(p - 1)$.

2) Selects two polynomials $f(x), g(x)$ randomly and secretly, whose orders are $(t - 1)$ and $(t_1 - 1)$ respectively.

3) Chooses α as primitive element of finite field Z_q.

Parameters set including (p, q, α) and $x_i, y_j \in_R Z_q[x]$, $(i = 1, 2, ..., n, j = 1, 2, ..., n_1)$, is public.

3.3   Generation of group key and secret pieces

We adopt the Shamir's secret sharing scheme[11], called SSS. Group secret key is produced by KAC, i.e., $(f(0) + g(0)) \mod q$ and group public key is $z = \alpha^{f(0)+g(0)} \mod p$.

Distribution of secret pieces: Being common member, i gains piece $f(x_i)$, and $z_i = \alpha^{\lambda_i f(x_i)} \mod p$. Otherwise, i.e., being a privilege member, $i \in G_1$ obtains pieces $\lambda_i f(x_i) + \mu_i g(y_{i,j}) \mod p$, where, λ_i, μ_i are SSS'public computable parameters [11] used to recover group secret key and $y_{,j}$ is namely the respective y_i introduced in section 3.2. KAC publish all z_i.

3.4   Generation of threshold group-signature

We might as well assume there are exactly t members taking part in, named 1,2,...,t. Suppose m is the message to be signed.

1) Generation and verification of individual signature   For any $i \in \{1, 2, ..., t\}$, i firstly selects $k_i \in Z_p^*$ secretly, and then computes $r_i = \alpha^{k_i} \mod p$, broadcasting r_i to all members. So each member can compute

$$r = \prod_{i=1}^{t} r_i \mod p$$

common member i, continues to compute $s_i = (f(x_i)\lambda_i h(m) - k_i r) \mod q$. And for privilege member i, he computes $s_i = (f(x_i)\lambda_i h(m) + g(y_{ij})\mu_i h(m) - k_i r) \mod q$, where, $h( )$ is some hash function. Finally, s_i is sent to SC.

SC can verify the validity of s_i. In fact, for any member i, the verification equation of SC is given as $\alpha^{s_i} r_i^r = z_i^{h(m)}$. If the equation holds true, s_i has been verified.

2) Combination of signature   If SC accepts all the individual signatures, it computes $s = (s_1 + ... + s_t) \mod q$ and outputs (r, s) as the group signature.

3.5   Verification and Traceability

Verification equation is given as: $\alpha^s r^r = z^{h(m)}$. Obviously, if satisfies the need for privilege threshold condition, we have results as follows according to SSS[12]

$$s = h(m)(\sum_{i=1}^{t} f(x_i)\lambda_i + \sum_{i=1}^{t_1} g(y_{ij})\mu_i) - r \sum_{i=1}^{t} k_i$$

$$= h(m)(f(0) + g(0)) - r \sum_{i=1}^{t} k_i$$

In view of $r = \prod_{i=1}^{t} r_i$ The verification equation mentioned above holds true. Otherwise, i.e., not satisfying those conditions, it is impossible to recover group secret key $f(0) + g(0)$ on the assumption that DLP is difficult to solve. Whenever any accredited agency want to investigate the identities of all member referring their individual signatures, the tracing procedure with assistance of SC is obvious.

3.6   Threshold group-signature scheme with several privilege subsets

Generic threshold group-signature problem is given in section 1. We just as well call such schemes $(t_1, n_1; ...; t_m, n_m; t, n)$-schemes, where, m is the number of privilege subsets and not smaller than 1.

$(t_1, n_1; t, n)$-schemes can be easily extended to $(t_1, n_1; ...; t_m, n_m; t, n)$-schemes. In fact, we may select $m + 1$ polynomials, $f(x), g_1(y_1), ..., g_m(y_m)$, and group secret key is $(\sum_{i=1}^{m} g_i(0) + f(0)) \mod q$, where any common member only obtain some piece $f(x_i)$ and any privilege member can gain knowledge of $f(x_i)$ and one respective $g_i(y_{ij})$, i.e., $\lambda_i f(x_i) + \mu_{ij} g_i(y_{ij})$, detail omitted.

3.7   Instance without the assistance of KAC

Similarly with [4], our schemes can do without the assistance of KAC. As a matter of fact, we can realize the schemes in such way, which each member becomes KAC of himself, i.e., selecting his public and secret key pair (x_i, y_i) (ElGamal type) by himself. The group public key is $y = \prod_{i=1}^{n} y_i$ Every member can distribute their own secret key pieces to privilege members in "double SSS "way and to members without privilege in common way. Other details are the same as the former sections.

4.   Threshold group-signature schemes with message recovery

In view of efficiency, signature scheme with message recovery is very attractive. In this section, we will give two $(t_1, n_1; ...; t_m, n_m; t, n)$-threshold group-signature schemes with message recovery. For simplicity, it is illustrated with $(t_1, n_1; t, n)$-threshold schemes all the same.

4.1   Generic threshold schemes of ElGamal type

Based on the discussion above, one may want to ask whether all of individual signatures of ElGamal type can be applied to construct threshold group-signature in such way. The answer is positive.

Firstly, the two individual signature schemes determined by

$$\begin{cases} r = \alpha^k \mod p \\ s = xh(m) \pm kr \mod q \end{cases} \tag{1}$$

and

$$\begin{cases} r = \alpha^k \mod p \\ s = xr \pm h(m) \mod q \end{cases} \tag{2}$$

Both of them can be used in the way above, where derivation of verification equation is omitted and concrete construction method similar with former sections. In fact, (1) is just the individual schemes used in section 3.

Noteworthily, [9, 12] proposed six variants of individual signature schemes with message recovery of ElGamal type. The two variants of them as follows can be used to construct our threshold group signature schemes with privilege subsets.

$$\begin{cases} r = R(m)\alpha^{-k} \mod p \\ s = -xr + k \mod q \end{cases} \tag{3}$$

and

$$\begin{cases} r = R(m)\alpha^{-k} \mod p \\ s = -x + kr \mod q \end{cases} \tag{4}$$

where, $R(\ )$ is redundancy function which is a reversible permutation(when replacing it with hash function, the two schemes are the same as common schemes, without message recovery.) and the respective verification equations are $R(m) = ry^r\alpha^s \mod p, R(m) = ry^{r^{-1}}\alpha^{sr^{-1}} \mod p$.

In section 4.2, taking the first variants above for example, we propose new $(t_1, n_1; t, n)$-threshold schemes.

4.2   Threshold schemes with message recovery

Initialization and distribution of secret key pieces are similar with section 3. We also assume there are exactly t members taking part in, named 1,2,...,t. Suppose m is the message to be signed. Procedure of signing is given as follows.

1) Generation and verification of individual signature For any $i \in \{1, 2, ..., t\}$, i firstly selects $k_i \in Z_p^*$ secretly, and then computes $r_i = R(m)\alpha^{-k_i}$ mod p, broadcasting r_i to all members. Therefore every member can compute the result

$$r = \frac{1}{R(m)^{t-1}} \prod_{i=1}^t r_i \mod p$$

For common member i, he or she continues to compute $s_i = (-f(x_i)\lambda_i r + k_i)$ mod q. And for privilege member i, he or she can compute $s_i = (-f(x_i)\lambda_i r$

$- g(y_{ij})\mu_i r + k_i)$ mod q, where, $h(\ )$ is some hash function. Finally, s_i is sent to SC.

For any member i, the verification equation of SC is given as: $r_i \alpha^{s_i} z_i^r = R(m)$. If the equation holds true, s_i has been verified.

2) Combination of signature If SC accepts all the individual signatures, it computes $s = (s_1 + ... + s_t)$ mod q and output (r, s) as the group signature. Obviously, the verification equation of group signature is: $r\alpha^s z^r = R(m)$.

As section 3, we can also construct the $(t_1, n_1; ...; t_m, n_m; t, n)$-threshold schemes with message recovery, named MR-$(t_1, n_1; ...; t_m, n_m; t, n)$-threshold schemes, which need not the assistance of KAC.

5.   Analysis

We take the scheme of section 3 called $(t_1, n_1; t, n)$−threshold schemes and the one of section 4 called $MR-(t_1, n_1; t, n)$−threshold schemes for examples. Because of restricted space, the proofs is omitted.

Theorem 5.1. *If each party involved in abides by rule of protocol, verification equations of the two schemes mentioned above, i.e., $\alpha^s r^r = z^{h(m)}$ and $r\alpha^s z^r = R(m)$, hold true.*

By all appearances, only SC may identify the members taking part in signing. So, our schemes satisfy the anonymity and do not need special authentication algorithm of identity, which is different greatly from [8]. In addition, our schemes have another good property, i.e. even the coalition of SC and some set of members ,which does not satisfy the threshold privilege condition, can not construct a valid signature because SC oneself can't see the secret group key and the first component of signature (i.e.,r)is produced by all members taking part in.

Finally, we study whether the two schemes are secure against forgery. We adopt "provable security" methodology to solve the problem. In other words, Lengthy research results have made ones be convinced of security against forgery of the two individual signature schemes (1) and (3), which may be seen in [9, 12]. Therefore, we may base our analysis on the assumption that the individual signature schemes (1) and (3) are secure. For the sake of simplicity, we prescribe that adversary is a probabilistic polynomial time(PPT) algorithm, which can corrupt any member of group.

We firstly prove that adversary, which can corrupt $(t - 1)$ members of G at most, see nothing from the interaction between she and honest members.

We mark $S(m, k) = (m, r, s)$ denoting random variable dominated by individual signature scheme (1) and $MS(m, k) = (m, r, s)$ by individual signature scheme with message recovery (3). Similarly, we may define random variables $GS(m, k_1, ..., k_t) = (m, \sigma, \tau)$, $GMS(m, k_1, ..., k_t) = (m, \sigma, \tau)$ according to group signature scheme of section 3 and the one of section 4 respectively, where assume the respective group secret key is the same as the secret key of individual scheme.

Theorem 5.2. *If the conditions above satisfied, then $S(m, k)$ and $GS(m, k_1, ..., k_t)$ are indistinguishable, so do MS and GMS.*

Secondly, we consider the interaction between adversary A and dishonest member, i.e. A tries to impersonate honest members. For the sake of simplicity, we assume A has corrupt $(t - 1)$ members out of all t participants. Take $(t_1, n_1; ...; t_m, n_m; t, n)$−threshold scheme for example and base our analysis on the assumption that the individual signature schemes

$$\begin{cases} r_i = \alpha^{k_i} \mod p \\ s_i = y_i h(m) - k_i r \mod q \end{cases} \tag{5}$$

is secure, where $r = r'r_i$ and r' is any extra public input.

Theorem 5.3. *Under the assumption above and conditions of theorem 3, $(t_1, n_1; ...; t_m, n_m; t, n)$-threshold scheme is secure against forgery.*

References

[1] Rivest,R.L., Shamir,A.and Adleman, L., A method for Obtaining Digital Signatures and Public-key Cryptosystem, Comm.ACM Vol.21(2),1978.

[2] ElGamal,L., A public Key Cryptosystem and a Signature Scheme Based on Discrete Logrithm, IEEE Trans. IT-31, 1985.

[3] Y.Desmedt and Y.Frankel, Shared generation of authenticators and signatures, CRYPTO'91, Springer-Verlag, 1992.

[4] L.Harn, Group-oriented (t,n)-threshold digital signature scheme based on discrete logarithms, IEE Proc. Computers and Digital Techniques, Vol.141, No.5, 1994.

[5] Jinn-Ke Jan, et al., A threshold signature scheme withstanding the conspiracy attack, Computer Communications, Vol.21, No.8, 1999.

[6] Wang Gui-lin and Qing Si-han, A Threshold Undeniable Signature Scheme Without a Trusted Party, Journal of Software, Vol.13,No.9, 2002.

[7] Kazuo Takaragi, et al., A Threshold Digital Signature Issuing Scheme without Secret Communication, {takara,kunihiko,takihasi}@sdl.Hitachi.co.jp,1997.

[8] Shi Yi and Feng Dengguo, The design and analysis of a new group of (t_j, t, n)−threshold group-signature scheme, ChinaCrypto'2000.

[9] Kaisa Nyberg, Rainer A.Rueppel, Message Recovery for Signature Schemes Based on the Discrete Logarithm Problem, EUROCRYPT'94, Springer-Verlag, 1994.

[10] Lidl, R. and Niederreiter, H., Introduction to Finite Fields and their Applications, London: Cambridge University Press,1986.

[11] Shamir A. How to share a secret. Communications of the ACM, 1979, 22(11).

[12] Giuseppe Ateniese and Breno de Medeiros, Efficient Group Signatures without Trapdoors, {ateniese,breno}cs.jhu.edu£¬2003.

A NEW GROUP SIGNATURE SCHEME WITH UNLIMITED GROUP SIZE

FU Xiaotong, XU Chunxiang
Information Security and Privacy Institute, National Key Lab of ISN
Xidian University, Xi'an, 710071, China
fxtt@163.com

Abstract   Group signature schemes are fundamental cryptographic tools that have many practical applications. We found there are some similarities between group signatures and proxy signatures. A new group signature scheme with unlimited group size is proposed based on the heuristic idea of proxy signature scheme with proxy signer's privacy protection. In this group signature scheme the group manager plays the role of the original signer, and the legal group members act as the proxy signers who delegate the original signer's signing power. The security properties of the new scheme are also discussed in the paper.

Keywords: group signature, proxy signature, privacy and anonymity, cryptographic protocols

Introduction

Group signature introduced by Chaum and Heyst [1] allows any member of a potential large group to sign on behalf of the group. It is required to be anonymous and unlinkable for anyone except the designated group manager who can co-relate signatures and reveal the identity of the actual signer. At the same time, no one (including the group manager) can misattribute a valid group signature. Group signatures are claimed to have many practical applications at present. The salient features—anonymity and unlinkability— of group signatures make them attractive for many specialized applications such as e-voting and e-auctions. This implied that proxy signatures with proxy signer privacy protection could be applied to construct group signature scheme with unlimited group size. A proxy signature scheme [2, 3, 4], which is a variation of ordinary digital signature scheme, enables a manager (called original signer) of a company to delegate his signing power to a reliable secretary (called proxy signer) who can sign on behalf of him. Proxy signature scheme with proxy signer privacy protection [5] is the signature that focuses on protecting the privacy of the proxy signer. Under this scheme anyone who only gets the proxy signatures cannot determine the actual identity of the proxy signer. This property can guarantee the proxy signers' security when they sign on sensitive information or they do not want to release their proxy identities to other third parties.

Mambo et al. [2] defined properties that a strong proxy signature scheme should provide:

1 **Strong Unforgeability:** Only the legitimate proxy signer can generate a valid proxy signature; even the original signer cannot.

2 **Verifiability:** Anyone can verify the signature and the signed message should conform to the delegation warrant.

3 **Strong Identifiability:** Anyone can determine the identity of the corresponding proxy signer either directly or indirectly (in this case, original signer's help is needed).

4 **Strong Undeniability:** The proxy signer cannot repudiate the signature that he ever generated.

A group signature scheme usually involves three entities: the group manager, group members, and the verifier, denoted by A, B and V respectively. The group manager is responsible for registering new user to the group, and only he can help a verifier to revoke the actual identity of the signer from the group signature in case of a dispute.

Definition1: In general, a *group signature scheme* is defined by a family of procedures [6]:

1 **SETUP**: A probabilistic algorithm, which generates the group-specific parameters.

2 **JOIN**: A prospective member executes this protocol, interacting with the group manager, to join the group. The new member's outputs are a membership certificate and the corresponding secret.

3 **SIGN**: A probabilistic algorithm outputs a group signature when given as input a message, the group public key, a membership certificate, and the associated membership secret.

4 **VERIFY**: A Boolean-valued algorithm used to test the authenticity of signatures generated from above step.

5 **OPEN**: An algorithm when given as input a message, a group signature on it, and other information needed, extracts the membership certificate used to issue the signature, and a non-interactive proof of the signature's authorship.

Definition2: The security properties that a group signature scheme must satisfy are as following [6]:

1 **Anonymity**: Given a valid signature, identifying the actual signer of the group is computationally hard for everyone but the group manager.

2 **Unforgeability**: Only legal members of the group can sign messages on behalf of the group; and other group members (even the group manager) cannot sign on behalf of a group member, which means they can not forge a valid group signature and successfully attribute it to another legal group member.

3 **Unlinkability**: Deciding whether two different group signatures produced by the same group member is computationally hard.

4 **Traceability**: The group manager is able to open a signature and identify the actual signer. Moreover, a signature signer cannot prevent the opening of a valid signature.

5 **Exculpability**: Neither a group member nor the group manager can sign on behalf of other group member.

6 **Coalition-Resistance**: A colluding sub set of group members cannot generate valid group signatures that cannot be traced.

In the following sections, this paper is organized as follows. In the second section we present a proxy signature scheme with proxy signer's privacy protection. In the third section based on the proxy signature scheme in section 2, a new group signature scheme with unlimited group size is proposed. In the fourth section we show the properties the new group signature scheme satisfy. In the fifth section the relations between group signatures and the proxy signatures are discussed. The last section is our conclusion.

1. Proxy signature with privacy protection

1.1 Notations

Through this paper, we define the system parameters in all schemes are as follows: Let p and q be two large prime numbers with $q|(p-1)$, and g be a generator of order q in the multiplicative group Z_q^*, $h(\cdot)$ be a cryptographic secure hash function. (x_A, y_A) is the secret-public key pair of user A. A user B's identity denoted by ID_B and we use $|ID_B|$ to denote the user B whose identity is ID_B, i.e. $|ID_B| = B$. Let $s = Sign(m, x_A)$ be a DLP-based signature on message m using private key x_A and $Verify(m, s, y_A)$ be the corresponding verification algorithm. Let "$||$" be the concatenation of two strings.

1.2   An improved proxy signature scheme

Identity blindness

In this step, the original signer A blinds all of his designated proxy signer's actual identities by giving every proxy signer in her valid proxy signer set $\{ID_B\}$ a new identity called proxy identity. Thus the proxy signer can sign on behalf of the original signer while hiding his actual identity to obtain privacy protection. For example, to blind the identity of a proxy signer B, A randomly chooses a number $k_B \in_R Z_q^*$, and computes $ID_P = h(k_B\|ID_B)$. A Sends ID_P to B. The hash value ID_P is the proxy identity of the proxy signer B. At the end of this step, original signer A records the tuple (ID_B, ID_P, k_B), for later use of anonymity revocation.

Delegating

Original signer A uses Schnorr signature scheme to delegate his signing power to proxy signer B whose proxy identity is ID_P. A chooses a random number $k_A \in_R Z_q^*$,

Computes $r_A = g^{k_A} \bmod p$

Computes $s_A = x_A h(ID_P\|r_A) + k_A \bmod p$

Then sends the ordinary Schnorr signature (r_A, s_A) to ID_P. s_A is the delegating key of A. ID_P verifies s_A by Schnorr signature verifying equation $g^{s_A} = y_A^{h(ID_P\|r_A)} r_A \bmod p$, he accepts (r_A, s_A) if the equation holds.

Proxy signing

$|ID_P|$, is actually B, produces his proxy key, denoted by x_P

Computes $x_P = s_A + k$

Compute $r = g^k \bmod p$

where $k \in_R Z_q^*$ is a random number.

Then $|ID_P|$ chooses one of the existing DLP-signature schemes to produce a proxy signature on massage m.

Compute $\sigma = Sign(m, x_p)$

the tuple $(m, \sigma, r_A, r_B, ID_P, r)$ is a valid proxy signature on behalf of A.

Verify

To verify a proxy signature, a verifier first computes the proxy public key y_P. Computes $y_P = y_A^{h(ID_P\|r_A)} r_A r \bmod p$

Then verify the received proxy signature as ordinary signature scheme to conform whether the equation $Ver(m, \sigma, y_P) = true$ holds or not.

Anonymity revocation

In case of a dispute, original signer can detect the actual proxy signer's identity ID_B from the recorded tuple (ID_B, ID_P, k_B) by ID_P in the proxy signature $(m, \sigma, r_A, r_B, ID_P, r)$.

2.   Group signature with unlimited group size

The proposed group signature scheme with unlimited group size contains five steps:

Step1: SETUP

When a user B wants to join a group G, he first sends a request message to the group manager A to start the protocol. B is required to convince A that he is a qualified member to join the group G. The authentication of group member can be achieved by ordinary authenticate schemes.

Step2: JOIN

- B generates a secret-public key pair (x, y), where $x \in_R Z_q^*$, $y = g^x \bmod p$, then sends the public key y together with his identity ID_B to A.

- A generates a random number $k_B \in_R Z_q^*$ and hashes it with ID_B. The hash value ID_P is the blind identity for B, i.e.
  Computes $ID_P = h(k_B \| ID_B)$.

- A then signs B's new public key y and blind identity ID_P together. The certificate also contains a warrant message w_P, which states that the blind identity ID_P is a legal member of group G. The certificate is (y, ID_P, w_P, r_A, s_A). To generate the certificate, A first chooses a random number $k_A \in_R Z_q^*$, and computes r_A, s_A
  Compute $r_A = g^{k_A} \bmod p$
  Compute $s_A = x_A h(y \| ID_P \| r_A) + k_A \bmod p$
  where x_A is the secret key of A, the corresponding public key is y_A.
  A records $(y, ID_P, w_P, ID_B, k_B)$ and sends (y, ID_P, w_P, r_A, s_A) to B.

Step3: SIGN

B uses his private key x to sign a message m on behalf of the group G. He creates the signature by using a DLP-based signature scheme:
Compute $\sigma = Sign(m, x)$.
The group signature is the tuple: $(m, \sigma, y, ID_P, w_P, r_A, s_A)$.

Step4: VERIFY

When verifying a signature created by a legal group member, the verifier first verifies the warrant w_P and the equation $g^{s_A} = y_A^{h(y \| ID_P \| w_P \| r_A)} \cdot r_A \bmod p$ to confirm the group manager's authentication of the group member. If above two verifies were both correct, the verifier verifies the received group signature σ by verifying-algorithm of the DLP-based signature scheme: $Veryfy(m, \sigma, y)$. If the value is true, the verifier accepts the group signature.

Step5: OPEN

In case of a dispute, the group signature protocol requires the group manager can identify the corresponding signer who produced the group signature by searching the database for the recorded tuple $(y, ID_P, w_P, ID_B, k_B)$ from the group signature tuple $(m, \sigma, y, ID_P, w_P, r_A, s_A)$.

A sends ID_B, k_B to the verifier.

The verifier computes $ID'_P = h(k_B\|ID_B)$,

Compares whether $ID'_P = ID_P$.

If they are the same, the verifier is convinced that the signer of the group signature is the group member B.

Thus, the actual identity of the group signer who produced the disputed signature has been revoked.

3. Properties analysis

1 **Unforgeability**:

In our protocol, obviously no one even the group manager can produce a legal group signature under the name of B. This is because the group manager and other signers or parties (either legal or illegal) do not have the secret key x. Nobody can produce a valid signature $\sigma = Sign(m, x)$ under the name of ID_P.

2 **Exculpability**:

The signature produced by a group member ID_P cannot be successfully attributed to another. This is because the group signature $(m, \sigma, y, ID_P, w_P, r_A, s_A)$ contains group manager's warrant w_P on the group member who signed the massage. At the same time, the group manager and a group member having no group member's secret key x could not generate signatures on behalf of other group members.

3 **Anonymity**:

Since the group signature $(m, \sigma, y, ID_P, w_P, r_A, s_A)$ does not contain any information directly related to the actual identity of B, no one can get anything useful to identify the actual signer from the group signature only. The group member's privacy is protected as proxy signer's privacy protection.

4 **Traceability**:

With the help of the group manager, $(y, ID_P, w_P, ID_B, k_B)$ can be detected from the information provided by the group signature. The verifier can know the actual identity of the signer in case of a dispute.

5 **Verifiability**:

The verifier can be convinced that a valid group member creates the group signature. This is because the public key y used in verifying-algorithm is certified by the group manager, and the warrant w_P contains the correlative information that states the blind identity ID_P in the group signature is the identity of a legal member of G. These two conditions hold only if the signature is actually created by a valid group member.

6 **Coalition-Resistance**:
By the use of the warrant w_P and the Unforgeability property, any group members cannot collude together to generate a signature and successfully claim it as a valid group signature signed by a legal group member B.

7 **Unlimited group size**:
The group manager does not have to determine how many members the group had before registering new members. This is because adding new group members have no impact on the group and the former members. The administrator can also register new group members at any time. As a result, the group size is unlimited.

4. Discussion

Proxy signatures with proxy signer's privacy protection and group signatures have the following similarities:

Authentication: In both schemes, the proxy signer (the group member) has to get the authorization from the original signer (the group manager) before he possesses the signing power.

Anonymity revocation: The signature created by any of the proxies (legal member in the same group) means the same to a verifier. If disputes do not occur, there is no need to identify which proxy (group member) created the signature.

Traceability: Both schemes should have identifiability (Traceability) and anonymity (Unlinkability).

Unforgeability: The valid signature belonging to a legal signer cannot be forged by anyone else.

As a result, the group signature scheme presented in this section actually transforms the proxy signer of a proxy signature scheme to the group member of a group signature scheme. The new scheme satisfies the group signature scheme's basic security properties better. At the same time the group size is improved to be unlimited, which is a good property required by many group signatures.

5. Conclusion

In this group signature scheme with unlimited group size, the group manager plays the role of the original signer, and the legal group member act as a proxy signer who delegates the original signer's signing power. The techniques developed in our group signature scheme are suitable for the applications in which the signer's anonymity and identifiability are required, such as e-commerce, e-voting and e-auction etc. They can also be used in electronics transaction [4] and mobile agent environment [5], so the proposed group signature scheme would have a very wide range of applications.

References

[1] D. Chaum, E. van Heyst. Group signatures. *EUROCRYPT'91*, Springer-Verlag, LNCS 547:257–265, 1991.

[2] M. Mambo, K. Usuda, and E. Okamoto. Proxy Signature: Delegation of the Power to Sign Messages. *IEICE Trans. Fundamentals*, E79-A(9):1338–1353, 1996.

[3] S. Kim, S. Park, and D. Won. Proxy signatures, revisited. *Proc. Of ICICS'97, International Conference on Information and Communications Security*, Springer-Verlag, LNCS 1334:223–232, 1997.

[4] H. Petersen, P. Horster. *Self-certified keys–concepts and applications*. Chapman & Hall, 102-116, 1997.

[5] K. Shum, V. Wei. A Strong Proxy Signature Scheme with Proxy Signer Privacy Protection. *Proc. of WETICE'02–Eleventh IEEE International Collaborative.*

[6] Ateniese, G. Tsudik. Some open issues and directions in group signature. In *Financial Crypto'99*, Springer-Verlag, LNCS 1648: 196–211, 1999.

[7] P. Kotzanikolaous, M. Burmcster, and V. Chrisskopoulos. Secure trans-actions with mobile agents in hostile environments. *Proc. ACISP*, LNCS 1841:289-297, 2000.

[8] B. Lee, H. Kim, and K. Kim. Strong proxy signature and its applications. *Proc of SCIS*, 603-608, 2001.

[9] G.B. Lee, H. Kim, and K. Kim. Strong proxy signature and its applications. *Proc of SCIS*, 603-608, 2001.

[10] Lijiang Yi, Guoqiang Bai, and Guozhen Xiao. Proxy multi-signature scheme. *Electronics Letters*, 36:527-528, 2000.

[11] Jan Camenisch, Markus Michels. A group signature scheme with improved efficiency. In *Proc. of ASIACRYPT '98*, Springer-Verlag, LNCS 1514, 1998.

[12] J. Camenisch, M. Stadler. Efficient group signature schemes for large groups. In *Proc.of CRYPTO'97*, Springer-Verlag, LNCS 1296, 1997.

[13] M. Bellare, D. Micciancio, and B. Warinschi. Foundations of Group Signatures: Formal Definitions, Simplified Requirements, and a Construction based on General Assumptions. In *Proc. Of EUROCRYPT '03*, Springer-Verlag, LNCS vol. 2656, 2003.

IDENTITY BASED SIGNATURE SCHEME BASED ON QUADRATIC RESIDUES

Weidong Qiu
Dept. of Communication Systems
University of Hagen
58084 Hagen, Germany
weidong.qiu@fernuni-hagen.de

Kefei Chen
Dept of Computer Science and Engineering Shanghai Jiaotong University
200030 Shanghai, China
kfchen@mail1.sjtu.edu.cn

Abstract   Identity based cryptosystems can greatly reduce the reliance on the current public key certificates which needed to be obtained in advance, by deriving public key from identification information such as an email address or a public IP address which can uniquely identify the entity. In this paper, we present a new identity based signature(IBS) scheme based on quadratic residue problem(IBS-QR). It is a combination of identity based and mediated cryptography. Furthermore, it can solve the public key revocation problem easily.

Keywords: identity based signature, quadratic residue, digital signature scheme

1.   Introduction

Nowadays, public key infrastructure(PKI) plays an important role in authenticating and providing binding between the entities and the corresponding public keys. It is a significant tool widely used in electronic commerce and secure communications areas. We notice that, under the current infrastructure, the certificate containing the signer's identity and the claimed public key has to be obtained in advance or has to be sent along with the signature before any authentication or verification of signatures takes place. The cost to establish and maintain a Certificate Authority(CA) is heavy. Furthermore, cross domain verification is complicated and some problems related with certification revocation list are still left unsettled like a "hot potato". As so far, many CAs have been set up but few of them succeeded. This is the main reason why individuals and organizations still keep caution when they plan to deploy PKI technology.

Unlike conventional PKI, there is no need to obtain public key in advance in identity based cryptosystems. Also, there is no need to establish an CA center. Public keys of the signers can be constructed with their public unique identification information such as email addresses etc. Consequently, the side effect brought by the cross domain verification and certification revocation problem can be eliminated naturally in identity based signature schemes. In general, identity based cryptosystems bring fresh air and new ways for constructing public cryptosystems.

The concept of identity based cryptosystems was firstly introduced by Shamir in 1984 [8]. Over the years, there is not so much development on this topic. Recently, Cocks proposed an identity based encryption scheme based on quadratic residue problem [3]. Boneh and Franklin developed another identity based encryption scheme BF-IBE based on Weil pairing [1]. Afterwards, a number of identity based cryptosystems have been presented within these two years [2, 6, 5, 9]. None of them except Cocks's scheme is based on quadratic residue problem. In Cocks's scheme, identity encryption based on quadratic residue problem was used to transfer keys of the symmetric key algorithm. The efficiency of the scheme is quite low for 128-bit symmetric encryption key needing 16K bytes of keying material or even double if the sender does not know she sends the square root of a or $-a$.

In this paper, we present a simple identity based signature scheme based on quadratic residue(IBS-QR). The signature scheme has the following advantages: the proposed scheme has a large signing space. Plaintexts are not limited only to the quadratic residues as normal Rabin like signature schemes. The new scheme is secure against chosen-ciphertext attacks by using a secure one way hash function during the signature generation. Furthermore, the new scheme is a combination of identity based and mediated cryptography similar to the IB-mRSA developed in [9]. Our scheme however is based on quadratic residue problem. The purpose of the combination is to eliminate the possible compromise of the private key during the signature generation. Compared with the scheme presented in [9], the key generation in our scheme is much simpler and more efficient. From the reducing computation point of view, this improvement is significant

The rest of the paper is organized as follows. Some basic notations is introduced in the next section. We give a detailed description of IBS-QR and analyze the security of the scheme in section 3. In section 4. we outline a possible application scenario of IDB-QR. Finally, we make some comparison and conclude our work in the last section.

2.   Notation and related theorem

In this section, we introduce some notations and related theorem which will be used in the scheme.

Let n be an integer and $Z_n^* = \{k \in Z_n \mid (k, n) = 1\}$ be the multiplicative group of the integer modulus n, and $a \in Z_n^*$. The integer a is said to be a quadratic residue modulo n, if there exists an $x \in Z_n^*$ such that $x^2 \equiv a \bmod n$. Otherwise a is called a quadratic non-residue modulo n. The set of all quadratic residues modulo n is denoted by Q_n and the set of all quadratic non-residue is denoted by $\overline{Q}_n$.

Let $n = pq$ be a product of two different odd primes. Then $a \in Z_n^*$ is a quadratic residue modulo n, if and only if $a \in Q_p$ and $a \in Q_q$.

The Jacobi symbol is a generalization of the Legendre symbol to an integer n which is odd but not necessarily prime. Let $n \geq 3$ be an odd integer with prime factorization $n = p_1^{e_1} p_2^{e_2} \cdots p_k^{e_k}$. Then the Jacobi symbol $\left(\frac{a}{n}\right)$ could be defined as follow.

$$\left(\frac{a}{n}\right) = \left(\frac{a}{p_1}\right)^{e_1} \left(\frac{a}{p_2}\right)^{e_2} \cdots \left(\frac{a}{p_k}\right)^{e_k}.$$

In addition, we define $J_n = \{a \in Z_n^* \mid \left(\frac{a}{n}\right) = 1\}$, and let $\tilde{Q}_n = J_n - Q_n$ denote the set of false square root of modulo n.

We get the following facts from [7]. If n is a composite and the factorization of n is unknown:

- There is no efficient procedure known for deciding whether or not $a(\in J_n)$ is a quadratic residue modulo n other than by guessing the answer.

- It is believed that the quadratic residue problem is as difficult as factoring integers, although no proof of this is known.

- Finding square roots of $a \in Q_n$ modulo $n = pq$ is computationally equivalent to factoring the modulus n.

Williams integer. A Williams integer is a composite integer of the form $n = pq$, where $p = 3 \bmod 8$ and $q = 7 \bmod 8$ are distinct primes.

Theorem 1. Let $x \in J_n$ and $n = pq$ be a Williams integer. Then

$$\left(x^{2d}\right) \equiv \begin{cases} x, & \text{if} \quad x \in Q_n \\ n - x, & \text{if} \quad x \in \tilde{Q}_n \end{cases}$$

where $d = (n - p - q + 5)/8$.

Proof. We know the fact that if $x \in Q_n$, then $x^{(n-p-q+5)/8} \bmod n$ is a square root of x modulo n. Therefore, we have the result $x^{2d} \equiv x \bmod n$.

Let us now consider the case where $x \in \tilde{Q}_n$. It follows from the definition of $\tilde{Q}_n$ that, $\left(\frac{x}{n}\right) = \left(\frac{x}{p}\right)\left(\frac{x}{q}\right) = 1$ and $\left(\frac{x}{p}\right) = \left(\frac{x}{q}\right) = -1$. In addition, because

$n = pq$ is a Willams integer, we can express p and q as the following forms: $p = 8x_1 + 3 = 4x'_1 + 3$, $q = 8x_2 + 7 = 4x'_2 + 3$. According to the properties of Legendre symbol, it follows that $\left(\frac{-1}{p}\right) = (-1)^{(p-1)/2} = (-1)^{2x'_1+1} = -1$, $\left(\frac{-1}{q}\right) = (-1)^{(q-1)/2} = (-1)^{2x'_2+1} = -1$. Then we get the fact that $\left(\frac{-x}{p}\right) = \left(\frac{-1}{p}\right)\left(\frac{x}{p}\right) = 1$, $\left(\frac{-x}{q}\right) = \left(\frac{-1}{q}\right)\left(\frac{x}{q}\right) = 1$. It means that $n - x$ (or $-x$) $\in Q_n$. As mentioned above, we know that $(n - x)^{2d} \equiv (n - x) \bmod n$. Now we can conclude that $x^{2d} \equiv (n - x)^{2d} \equiv (n - x) \bmod n$ when $x \in \tilde{Q}_n$.

3. Identity based signature scheme based on quadratic residue problem(IBS-QR)

The basic idea behind IBS-QR scheme is the use of user's public unique identity such as email address etc., for deriving user's public key. All user within a system share a single common modulus n. The main feature of identity based signature is that there is no need to issue certificates which bind the identity of the certificate's holder and the corresponding public key. Revocation problem can be solved easily in identity based signature scheme in contrast to the complexity in current PKI systems.

It is obviously insecure when all users share a same single modulus within the system. In RSA-like public key systems, sharing a single modulus subjects to a vulnerability of factoring the modulus and consequently facilitates computing the other user's private key. Similarly, when sharing a single common modulus in a quadratic residue based identity signature scheme, the private keys of all users are the same. We approach the problem by introducing a mediated signer(MS) and split the private keys of users into two parts which are hold by the user and the MS separately so that no one possesses a complete private key. The same method was adopted in Ding's Identity based system [9]. Now we turn to the setup of the signature scheme.

The initialization work can be done by a trusted authority(TA). TA selects a Williams integer n, where $n = pq$, and p, q are primes near in length, as the public key of all system users. The system private key, (p, q, d) where $d = (n - p - q + 5)/8$, is known only to the TA.

For each user, TA selects a random $r \bmod n$(unique to each user) as the user-part private key. Then TA computes $R = (d - r) \bmod n$ as the MS-part private key. MS collects all MS-part keys of all users and the users generate signatures co-operating with the MS.

Let $h(x)$ be a collision resistant hash function.

3.1   IBS-QR signing

1. Signature generation Suppose Alice be the signer, MS be the mediated signer(MS), Bob be the verifier and the message to be signed be m. In our case, we use the user's email address as the public identity(id). The signature generation process is shown in figure 1 and figure 2.

Signing: Alice

- Compute $h(m||id), h(id)$
- Test whether $\gcd(h(m||id), n) = 1$ or not
      If this is the case, the message m should be changed
- Choose a as follows:

$$a = \begin{cases} 0, & \text{if} & \left(\frac{h(m||id)}{n}\right) = 1 \\ 1, & \text{if} & \left(\frac{h(m||id)}{n}\right) = -1 \end{cases}$$

    so that $2^a h(m||id) \bmod n \in J_n$
- Compute $s' \equiv (2^a h(m||id))^r \bmod n$
- Send the temporary result $s', a, m, id, h = h(s'||a||m||id)$ to the MS

Figure 1.   Signing process 1

Signing: MS

- After receiving s', a, m, id, verify the hash value h
- Choose the corresponding key R according to Alice's identity
- Compute the final signature $s \equiv s' \{2^a h(m||id)\}^R$
-                            $\equiv \{2^a h(m||id)\}^d \bmod n$
- Send the final signature result (s, a) back to Alice

Figure 2.   Signing process 2

During the signature generation process, the message should be changed if $\gcd(h(m||id), n) \neq 1$. Otherwise, it is possible to reveal the value of p or q^1. The probability of $(h(m||id), n) \neq 1$ is $\frac{1}{p} + \frac{1}{q} - \frac{1}{n}$, which can be ignored if n is large enough in length.

According to the literature [4], even without knowledge of the factoring of the modulus n, the Jacobi symbol can be calculated.

Furthermore, we assume that the connection between MS and Alice is secure, for instance, using secure mechanisms such as VPN or sharing encryption key etc.

2. Signature verification After receiving the pair $(m, (s, a))$, the verifier, Bob, calculates $h(m||id)$ using the Alice's email address which is already widely known. Then Bob tests the equation $s^2 2^{-a} \equiv h(m||id) \bmod n$ or

$s^2 2^{-a} \equiv n - h(m||id) \bmod n$. Bob accepts the signature only if one of the two equations is true.

Theorem 2. With the choice of the message space and the integer a, we have $2^a h(m||id) \bmod n \in J_n$.

Proof. With the choice of the message space and the integer a, the Jacobi symbol $(\frac{h(m||id)}{n})$ equals to 1 or -1. Since n is a Williams integer, so 2 is a quadratic non-residue modulo n, which means $(\frac{2}{n}) = -1$. In addition, it is obvious that $\gcd(2^a h(m||id), n) = 1$. Therefore, the following conclusion can be easily obtained: $(\frac{2^a h(m||id)}{n}) = 1$, which leads to the conclusion, $2^a h(m||id) \bmod n \in J_n$.

Theorem 3. If all parties involved in the signature generation carry out the protocol properly, then verifiers can make sure the signature is a valid one.

Proof. Let (s, a) be a right signature of message m. By Theorem 2, with the choice of the message space and the integer a, $2^a h(m||id) \bmod n$ belongs to J_n. In addition, n is a Williams integer, we know the fact from Theorem 1 that

$$s^2 2^{-a} \equiv \{(2^a h(m||id))^d\}^2 2^{-a} \equiv$$

$$(2^a h(m||id))^{2d} \, 2^{-a} \equiv \begin{cases} h(m||id), & \text{if } 2^a h(m||id) \in Q_n \\ n - h(m||id), & \text{if } 2^a h(m||id) \in \tilde{Q}_n \end{cases}$$

Now we can conclude that, the signature verification process can make sure the validity of the signatures generated by IBS-QR.

3.2   Security Analysis of IBS-QR

Let n be a large composite with unknown factorization. Computing the square roots modulo n is believed to be as difficult as factoring large integer [7]. We study the security of IBS-QR on the assumption that the modulus n can not be factorized.

For using hash function in the scheme, IBS-QR is chosen-ciphertext attack resistant: given x, one can compute $y = x^2 \bmod n$, but finding z satisfying $y = h(z)$ equals to computing the inverse of hash function. Given z, one can compute $y = h(z)$, but finding x is a quadratic residue problem. Furthermore, even if one has obtained a triple (x_0, y_0, z_0) where $x_0^2 \equiv y_0 \bmod n$, and $y_0 = h(z_0)$, it is difficult to find another triple (x_1, y_1, z_1) satisfying $x_1^2 \equiv y_1 \bmod n$, $y_1 = h(z_1)$. Otherwise, either $x_1 = x_0, y_1 = y_0$, but $z_1 \neq z_0$, which means that $h(x)$ is not a secure hash function, or $x_1^2 \equiv x_0^2 \equiv y \bmod n$ but $x_1 \neq x_0 \bmod n$, which means that $x_1 + x_0$ or $x_1 - x_0$ is one of the prime factors of modulus n.

This is impossible according to the assumption on the difficulty of factorizing the modulus.

For splitting the private key into two parts, no party owns a complete private/public key pair. This protects the systems from revealing information of the modulus. The mediated signer, since only knows one part of the private, has no ability of signing the messages on behalf of the system users. Even when the mediated signer was compromised, the knowledge of all different R does not enable the attacker to factorize the modulus or sign messages in the name of the system users because the private key part of the users will never transmitted to the mediated signer in plain text and the attacker will not be able to compute the r-part from the former signing process view. The system will be broken only under the case that the attacker compromised the mediated signer and colluded with at least one of the users from the system. Since the r-part of the private key is only known to the owner, no one else in the system can initiate a signing process in the name of the owner.

A possible attack is that Alice initiates a valid signature process with MS, and obtains a valid signature on m. Afterwards, she claims that this signature is from someone else, say, Bob. We would say this attack will not succeed because we have combined the signer's identity into the signature. When the receiver verifies the signature, Bob's identity($h(m||id_{Bob})$) instead of Alice's($h(m|| id_{Alice})$) will be used, and that will result in a failed signature verification.

Another possible forgery attack is as follows. Alice has the signature $(m, (s, a))$ signed by MS and herself. Then, she calculates the multiplicative inverse s^{-1} from $s \times s^{-1} \equiv 1 \mod n$.

Signing by Alice:

(1) compute $t \equiv 2^{a'} h(m'||id_b)^r \times 2^a h(m||id_a)^r \mod n$

(2) send the temporary result t, $2^{a'} h(m'||id_b) \times 2^a h(m||id_a)$ to the MS

Signing by the MS:

(1) select the corresponding key R according to Alice's identity

(2) compute the signature $s'' \equiv t' \times \{2^{a'} h(m'||id_b) \times 2^a h(m||id_a)\}^R$
$\equiv \{2^{a'} h(m' ||id_b) \times 2^a h(m||id_a)\}^d \mod n$

(3) send the result (s'', a') back to Alice

After receiving the pair (s'', a'), Alice calculates the final signature $s' \equiv s'' \times s^{-1} \equiv 2^{a'} h(m'||id_b)^d \mod n$. Then, the forged signature for m' is (s', a'). But this attack does not work in our system because the MS will check the hash value based on the tuple s', a, m, id.

The IBS-QR scheme is much more stronger if we assume that the connection between the MS and users are secure, say, using authentication and encryption.

4.   Practical aspects

We outline here one of the possible practical application scenarios to which IBS-QR scheme may be applied.

Let's consider the case when an organization, which has dozens of branches and thousands of employees all over the world with frequent flow of personnel, wants to deploy an internal signature system. If a conventional PKI architecture based on digital certificates system was adopted, the management cost of publishing certificates will be heavy. Even worse, because of the frequent personnel flow, the certification revocation list(CRL) will be very difficult and inefficient to maintain. Opposed to the conventional one, it is much convenient in IBS-QR scheme to revoke any private key–after the revocation, the MS simply rejects the signing request from that revoked user immediately–there is no delay.

The occurrence of a mediated signer will not be a barrier to the usage of the system. Especially when this system is adopted by an organization as internal signature system–it is not difficult to set up and maintain such an on-line server as mediated signer center.

As for computation cost aspect, compared with the normal signature schemes based on quadratic residue problem, $h(id)$ times modulus multiplication instead of two has to be calculated. However, on the other hand, there are less modulus multiplication in the signer's side compared with most signature schemes.

5.   Comparison and conclusion

We presented an identity based signature IBS-QR based on quadratic residue problem. The advantages of IBS-QR are that there is no need to publish public key and revocation is quite simple. These features are completely distinguished from the conventional way. The verifier can construct the public key of the signer through the public available information. We list some advantages of the presented scheme as follows.

- There is no need to establish a or more Certificate Authorities to issue certificates which bind the identities and the corresponding public keys.

- There is no cross verification problem.

- The revocation is quite simple. Under conventional PKI infrastructure, CRL (certificate revocation list) is issued to withdraw certificates and withdraw the public keys contained in these certificates as well. The problem is that whichever way to publish the CRL, there is delay. In IBS-QR scheme, there is no delay-when a public key is withdrawn, the MS simply stop providing the signature service for the corresponding entity. There is no delay.

The similar architecture has been adopted also both in BF-IBE and IB-mRSA while different cryptographic primitives were used. BF-IBE used the elliptic curve primitive and RSA was used in IB-mRSA. IBS-QR is based on the quadratic residue problem. All these schemes need a mediated center, or may be called as trusted thirty party. IBS-QR provides the same revocation mechanism as in IB-mRSA. While BF-IBF provides only a periodic revocation instead of a fine grained revocation as in IBS-QR and IB-mRSA. IBS-QR has almost the same performance as IB-mRSA, but better than BF-IBE as pointed by [9]. Compared with IB-mRSA, IBS-QR provides much more balance between signing cost and verifying cost. In addition, IBS-QR has much higher efficiency in key generation compared with BF-IBE and IB-mRSA. In IB-mRSA scheme, the key generation process is the same as normal RSA like algorithms. Modular inversion operations have to be done towards getting the private keys. While in IBS-QR, the only operation needed is to choose a random integer.

The use of mediated signer makes the system remaining secure while sharing a single modulus among all system users. For splitting the private key, the system as a whole is unforgeable and undeniable.

Furthermore, the signing message space is now not limited to the quadratic residues modulo n.

We try to construct signature scheme from a new angle and facilitate the using of public key system. It still remains uncertain whether an on-line mediated signer can run efficiently or not. Some testing implementation, if possible, should be conducted in the near future. In addition, the verifying process is more expensive compared with normal quadratic residue problem based signature schemes(only two modulus multiplication needed). Further investigation should be continued on these aspects.

Notes

1. As we know, $n = pq$. p and q are the only factor of n. If $\gcd(h(m||id), n) \neq 1$, then $h(m||id)$ must be the form of xp or $x'q$, where $x \in q - 1$ and $x' \in p - 1$. So it is much easier to get the factor p or q

References

[1] D. Boneh and M. Franklin. Identity based encryption from the weil pairing. In *Advances in Cryptology - CRYPTO '2001*, volume 2139 of *Lecture Notes in Computer Science*, pages 213–229. Springer Verlag, 2001.

[2] D. Boneh, B. Lynn, and H. Shacham. Short signatures from the weil pairing. In *Advances in Cryptology - ASIACRYPT '2001*, volume 2248 of *Lecture Notes in Computer Science*, pages 514–532. Springer Verlag, 2001.

[3] C. Cocks. An identity based encryption scheme base on quadratic residues. In *Cryptography and Coding*, volume 2260 of *Lecture Notes in Computer Science*, pages 360–364. Springer Verlag, 2001.

[4] H. Cohen. A course in computational algebraic number theory. volume 138. Springer-Verlag graduate texts in mathematics, 1993.

[5] F.Hess. Efficient identity based signature schemes based on pairings. In *Selected Areas in Cryptography*, volume 2595 of *Lecture Notes in Computer Science*, pages 310–324. Springer Verlag, 2003.

[6] J.C.Cha and J.H.Cheon. An identity-based signature from gap diffie-hellman groups. In *PKC2003*, volume 2567 of *Lecture Notes in Computer Science*, pages 18–31. Springer Verlag, 2003.

[7] Alfred J. Menezes, Paul C. van Oorschot, and Scott A. Vanstone. *Handbook of applied cryptography*. CRC Press series on discrete mathematics and its applications. CRC Press, 1997. ISBN 0-8493-8523-7.

[8] A. Shamir. Identity based cryptosystems and signature aschemes. In *Advances in Cryptology - CRYPTO '84*, volume 196 of *Lecture Notes in Computer Science*, pages 47–53. Springer Verlag, 1984.

[9] X.Ding and G. Tsudik. Simple identity-based cryptography with mediated rsa. In *Topics in Cryptology – CT-RSA 2003*, volume 2612 of *Lecture Notes in Computer Science*, pages 193–211. Springer Verlag, 2003.

A NEW DIGITAL SIGNATURE SCHEME BASED ON FACTORING AND DISCRETE LOGARITHMS*

Shimin Wei

Department of Computer Science & Technique
Huaibei Coal Normal College, Huaibei, 235000, China
weism02@yahoo.com.cn

Abstract   In 1994, He and Kiesler proposed a digital signature scheme which was claimed to be based on the factoring problem and the discrete logarithms problem. This paper shows that any attacker can forge the signature of He-Kiesler scheme without solving any hard problem. A new digital signature scheme is proposed, the security of which is based on both factorization problem and discrete logarithms problem.

Keywords: cryptography, digital signature, factoring, discrete logarithms, security

1.   Introduction

In 1994, Harn [1], He and Kiesler [2] proposed digital signature schemes based on two hard problems-the factoring problem and the discrete logarithms problem. Since then, many digital signature schemes based on these two hard problems were proposed [3, 4, 5, 6]. Unfortunately, most of them have shown to be insecure [7, 8, 9, 10]. For example, in 1995, Harn [7] showed that one can break the He-Kiesler scheme [2] if one has the ability to solve the factorization. At the same year, Lee and Hwang [8] showed that if one has the ability to solve the discrete logarithms, one can break the He-Kiesler scheme. This paper shows that any attacker can forge the signature of He-Kiesler scheme without solving any hard problem. Instead, we propose a new digital signature scheme, which is secure if one cannot solve the factorization problem and the discrete logarithms problem simultaneously.

* This work was supported in part by the Natural Science Foundation of China under Grant 60172015 and the Natural Science Foundation of Anhui Province in China under Grant 03042204.

2.   He-Kiesler scheme and a simple attack

2.1   He-Kiesler scheme

Let p be a large prime such that $p - 1$ has two large prime factors p_1 and q_1. Let $n = p_1 q_1$, and let g be a primitive element, or an element of large order, of $GF(q)$. Note that if a common p is used by all users, the two factors of n must be kept secret from every user (actually these two factors will never be used by anyone, and thus can be discarded once n is produced).

Any user A has a secret key x_1 $(1 < x_1 < n)$ such that $\gcd(x_1, p - 1) = 1$. From x_1 constructed the quadratic residue $x = x_1^2 \bmod (p-1)$ and corresponding public key $y = g^{x^2} \bmod p$. To sign a message m, A does the following

1) Randomly chooses an integer t_1 $(1 < t_1 < n)$ such that $\gcd(t_1, p - 1) = 1$, and calculates $t = t_1^2 \bmod (p - 1)$

2) Computes $c = x_1 t_1 \bmod (p - 1)$

3) Computes $r = g^{t^2} \bmod p$ and makes sure that $r \neq 1$

4) Finds s such that

$$m = xr + ts \bmod (p - 1) \tag{1}$$

5) Sends $Sig(m) = (r, s, c)$ as the signature.

To verify that (r, s, c) is a valid signature of m, one simply checks the identity

$$g^{m^2} \equiv z^{r^2} r^{s^2} g^{2rsc^2} \bmod p$$

2.2   A simple attack

Let (r, s, c) be a signature of a known message m. From (1) we have

$$mx = x^2 r + stx \bmod (p - 1).$$

Since $c^2 = tx \bmod (p - 1)$, the attacker obtains the second-order equation as

$$rx^2 - mx + sc^2 = 0 \bmod (p - 1) \tag{2}$$

Assume that (r', s', c') is a signature of a known message m' such that $mr' - m'r \neq 0 \bmod (p-1)$. Then the attacker obtains another second-order equation as

$$r'x^2 - m'x + s'c'^2 = 0 \bmod (p - 1) \tag{3}$$

Now the attacker can easily obtain x and x^2 by solving (2) and (3), and obtain t from x by solving (1).

Although the attacker cannot calculate the secret key x_1 from the signing key x, he can forge the signature of the signer A as follows. To sign a forged message m'', he chooses an arbitrary signature (r, s, c) and finds a s' such that the following equation

$$m'' = rx + ts''\mathrm{mod}(p - 1)$$

is satisfied. The signature of the message m'' is (r, s'', c).

3. Modified He-Kiesler Signature Scheme

3.1 The new scheme

Let p be a large prime such that $p - 1$ has two large prime factors p_1 and q_1. Let $n = p_1 q_1$, and let g be a primitive element of $GF(q)$. User A has a secret key x ($1 < x < n$) such that $\gcd(x, p - 1) = 1$. The corresponding public key is $y = g^{x^2} \bmod p$. To sign a message m, A does the following

1) Randomly chooses an integer t ($1 < t < n$) such that $\gcd(t, p - 1) = 1$,

2) Computes $r_1 = g^{t^2} \bmod p$ and makes $r_2 = g^{t^{-2}} \bmod p$ and makes sure that $r_1 \neq 1$

3) Finds s such that

$$mt^{-1} = xr_1 + ts^2 \bmod (p - 1) \tag{4}$$

4) Sends $Sig(m) = (r_1, r_2, s)$ as the signature.

To verify that (r_1, r_2, s) is a valid signature of m, one checks the identity

$$r_1^{s^4} r_2^{m^2} = y^{r^2} g^{2ms^2} \tag{5}$$

Now we show that if the signer follows the above protocol, the recipient always accepts the signature. From (4)

$$x^2 r_1^2 = m^2 t^{-2} + t^2 s^4 - 2ms^2 \bmod (p - 1)$$

so that

$$m^2 t^{-2} + t^2 s^4 = x^2 r_1^2 + 2ms^2 \bmod (p - 1)$$

hence

$$
\begin{aligned}
r_1^{s^4} r_2^{m^2} &= g^{t^2 s^4} g^{m^2 t^{-2}} = g^{t^2 s^4 + m^2 t^{-2}} \\
&= g^{x^2 r_1^2 + 2ms^2} = g^{x^2 r_1^2} g^{2ms^2} = y^{r^2} g^{2ms^2} \bmod p.
\end{aligned}
$$

Therefore, (5) holds.

In the He-Kiesler scheme, the signer does not need to know how to factor $p - 1$, thus it is possible for every user to employ the same modulus p, where p is the modulus commonly. But in the new scheme, every user must employ a different modulus p, and know how to factor $p - 1$.

In (4) and (5) of the new scheme, If m is replaced by $h(r_1, r_2, m)$, where h is a collision-free one-way hash function, the new scheme is more secure.

3.2   The security of the new scheme

As to the security of the new scheme, note the following.

(1) To recover x from public key $y = g^{x^2} \bmod p$ it is necessary to compute both the discrete logarithm of y modulo p (obtain x^2), and the square root of x^2 modulo n to obtain x.

(2) To recover x from (4) it is necessary to compute both the discrete logarithm of r_1 modulo p (obtain t^2), and the square root of t^2 modulo n to obtain t^2, or to compute both the discrete logarithm of r_2 modulo p (obtain t^{-2}), and the square root of t^{-2} modulo n to obtain t^{-1}.

(3) Even if an attacker has the ability to compute the discrete logarithms modulo a large prime number p, he can recover x^2, t^2, t^{-2} from y, r_1, r_2, respectively. To forge the signature of for any message m', the attacker must finds s' such that

$$m't^{-1} = xr_1 + ts'^2 \bmod (p - 1)$$

so that

$$x^2 r_1^2 = m'^2 t^{-2} + t^2 s'^4 - 2m's'^2 \bmod (p - 1)$$

To solve s' from above equation he must be face with another difficult problem: factoring $p - 1$.

(4) Assume that a signature (r_1, r_2, s) of a known message m is given. It is then possible for the attacker to establish the following equation as

$$mt^{-2} - s^2 = xr_1 t^{-1} \bmod (p - 1)$$

To sign a forged message m', he finds an s' such that the following equation

$$m't^{-2} - s'^2 = xr_1 t^{-1} \bmod (p - 1)$$

From above two equations it follows that

$$s'^2 = s^2 - mt^{-2} - m't^{-2} \bmod (p - 1)$$

. To recover s' it is necessary compute both the discrete logarithm of r_2 modulo p (obtain t^{-2}), and the square root of s'^2 modulo n to obtain s'.

4. Conclusion

We have shown that any attacker can forge the signature of He-Kiesler scheme without solving any hard problem. A modified He-Kiesler signature scheme is proposed, the security of which is based on both factorization problem and discrete logarithms problem.

References

[1] HARN, L.: Public-key cryptosystem design based on factoring and discrete logarithms, IEE Proc.-Computers and Digital Techniques, 1994, 141, (3), pp. 193-195

[2] HE, J., and KIESLER, T.: Enhancing the security of ElGamal's signature scheme, IEE Proc. -Computers and Digital Techniques, 1994, 141, (4), pp. 249-252

[3] HE, W. H.: Digital signature scheme based on factoring and discrete logarithms, Electron. Letters, 2001, 37, (4), pp. 220-222

[4] LEE, N. Y., and HWANG, T.: Modified Harn signature scheme based on factoring and discrete logarithms, IEE Proc.-Computers and Digital Techniques, 1996, 143, (3), pp. 196-198

[5] SHAO, Z.: Signature schemes based on factoring and discrete logarithms, IEE Proc.-Computers and Digital Techniques, 1998, 145, (1), pp. 33-36

[6] LAIH, C. S., and Kuo, W. C.: New signature schemes based on factoring and discrete logarithms, IEICE Trans. Fund, 1997, E80-A, (1), pp. 46-53

[7] HARN, L.: Comment: Enhancing the security of ElGamal's signature scheme, IEE Proc. -Computers and Digital Techniques, 1995, 142, (5), pp. 376

[8] LEE, N. Y., and HWANG, T.: The security of He and Kiesler's signature schemes, IEE Proc.-Computers and Digital Techniques, 1995, 142, (5), pp. 370-372

[9] LI, J., and XIAO, G.: Remarks on new signature scheme based on two hard problems, Electron. Lett., 1998, 34,(25), pp. 2401

[10] LEE, N. Y.: Security of Shao's signature schemes based on factoring and discrete logarithms, IEE Proc.-Computers and Digital Techniques, 1999, 146, (2), pp. 119-121

NEW TRANSITIVE SIGNATURE SCHEME BASED ON DISCRETED LOGARITHM PROBLEM*

Zichen Li, Juanmei Zhang

Dept of Computer Sci. & Tech., Jiaozuo Institute of Technology, Jiaozuo 454159, China

Dong Zheng

Dept of Computer Sci. & Eng., Shanghai Jiaotong University, Shanghai 200030, China

Abstract   In this paper, we will propose an new transitive digital signature scheme, which is generalized realization based on discreted logarithm problem. We also show that our new scheme is proven transitively unforgeable under adaptive chosen-message attack assuming discreted logarithm problem is hard.

Keywords: digital signature, transitive digital signature, cryptography

Introduction

Recently, Micali and Rivest first introduced the concept of transitive digital signature, which is that of dynamically building an authenticated graph, edge by edge[MicaRiv]. Informally, this is a way to digitally sign vertices and edge of a transitively closed graphed G, so as to guarantee the properties of Transitive and Unforgeable.

These new features are the algebric properties, and have many interesting applications. For a undirected example, where the graph represents a set of administrative domains. The nodes represent computers and an undirected edge means that u and v are in the same administrative domain. Again, it is clear that if u and v are in the same administrative domain and v and w are in the same administrative domain, the u and w are in the same administrative domain. According to the above concept, Bellare and Neven designed some novel realization of transitive signature scheme based on factoring and RSA[BellNev]. And these schemes are proven transitively unforgeable under adaptive chosen message attck assuming that factor is hard.

*Project Supported National Natural Science Foundation of China(No.90104032).

In this paper, according to the formal definition of transitive signature, described by Bellare and Neven in [BellNev], we will design an new transitive signature scheme based on the discrete logarithm problem. The new scheme is very different with Micali and Rivest's scheme.

1.   Definitions

In this paper, all graphs are undirected. A graph $G^* = (V^*, E^*)$ is the *transitively closed* if for all nodes $i, j, k \in V^*$ such that $\{i, j \in E^*\}$ and $\{j, k\} \in E^*$, it also holds that $\{i, k \in E^*\}$. In other words, the *transitive closure* graph G^*, E^* of $G = (V, E)$ is defined to have $V^* = V$ and to have an edge i, j in E^* if and only if there is a path (of length zero or greater) from i to j in G.

Transitive signature scheme:

A transitive signature scheme **TS = (TKG, TSign, TVf, Comp)** is specified by four polynomial-time algorithms, and the functionality is followings:

- The randomized key generation algorithm **TKG** takes input 1^k, where $k \in 1, 2, ...$, is the security parameter, and returns a pair (tpk, tsk) consisting of a public key and mathcing secret key.

- The signing algorthm **TSin** takes input the secret key tsk and nodes $i, j \in 1, 2, ...$, and returns a value called an original signature of edge $\{i, j\}$ relative to tsk.

- The deterministic verification alogrithm **TVf**, given tpk, nodes $i, j \in 1, 2, ...$, and a candidate signature σ, returns either 1 or 0. In the former case, we say that σ is a valid signature of edge $\{i, j\}$ relative to tpk.

- The deterministic compostion algorithm **Comp** takes tpk, nodes $i, j, k \in 1, 2, ...$, and values σ_1, σ_2 to return either a value σ or a symbol $\perp$ to indicate failure.

Correctness of transitive signature scheme:

In practice, it is desirable to allow users to name nodes via whatever identifiers they choose, but these names can always be encoded as intergers, so we make the simplify assumption that the nodes of the graph are positive intergers.

$(tpk, tsk) \overset{R}{\leftarrow} \textbf{TKG}(1^k)$

$S \leftarrow \emptyset$; **Legit** $\leftarrow true$; **NotOK** $\leftarrow false$

Run: A with its oracles until it halts, replaying to its oracle queries as follow:

If A makes **Tsign** quer i, j then

If $i = j$ then **Legiet**$\leftarrow false$

Else

Let σ be the output of the **TSign** oracle and let $S \leftarrow S \cup \{(\{i, j\}, \sigma)\}$

If **TVf**$(tpk, i, j, \sigma) = 0$ then **NotOK** $\leftarrow true$

If A makes **Comp** query $i, j, k, \sigma_1, \sigma_2$ then
    If $(\{i, j\}) \notin S]$ or i, j, k are not all distinct, then **Legit** $\leftarrow false$
    Else
    Let σ be the output of the **Comp** oracle and let $S \leftarrow S \cup \{(\{i, j\}, \sigma)\}$
    Let $\tau \leftarrow$ **Tsign**(tsk, i, k)
    If $[(\sigma \neq \tau)$ or **TVf**$(\mathbf{tpk, i, k}, \sigma) = 0]$ then **NotOK** $\leftarrow true$
When A halts, output $(\textbf{Legit} \wedge \textbf{NotOk})$

According to the above steps, the meaning of the correctness of a transitive signature schemem is that it is impossible for every alogarithm A to make a legitimate query and return a signature differs from the original one.

Definition 1 We say the transitive signature scheme **TS=(TKG, Tsign, TVf, Comp)** is correct if for every alogarithm A and every k, the output of the experiment of Figure 1 is true with probability zero.

The experiment computes a bollean **Legit** which is set to $false$ if A ever makes an illegitimate query. It also compute a boolean **NotOK** which is set to $true$ if a signature returned by the composition algorithm differs from the original one. To win , A must stay legitimate (meaning **Legit**=true) but violate correctness (**NotOK**=$true$) The experiment returns $true$ with probability zero.

Security of transitive signature scheme:
    For transitive signature scheme **TS=(TKG, TSign, TVf, Comp)** with the security $k \in N$, $\mathbf{Exp}_{TS,F}(k)$ denotes the attacked experiment done by adversary F. $\mathbf{Exp}_{TS,F}(k)$ returns 1 if and only if $F$is successful in its attack on this scheme. The experiment begins by runing **TKG** in input 1^k to get keys (tpk, tsk). If we are in the random oracle model, it also chooses the appropriate hash functions at random. It then runs F, providing this adversary with input tpk and oracle access to the function **TSign**$(tsk, ., .)$.The oracle is assumed to maintain state or toss coins as needed. Eventually, F will output $\{i, j' \in N\}$ and some value σ'. Let E be the set of all edges $\{a, b\}$ such that F made oracle query a, b, and let V be the set of all intergers a such that a is adjacent to some edge E. We say that F wins if σ' is a valid signature of $\{i', j'\}$ relative to tpk, but edge $\{i', j'\}$ is not in the transitive closure G^* of graph $G = (V, E)$.

The experiment returns 1 if F wins and 0 otherwise. The advantage of F in its attack on **TS** is the function $\mathbf{Adv}_{TS,F}(k)$ defined by

$$\mathbf{Adv}_{TS,F}(k) = Pr[\mathbf{Exp}_{TS,F}(k)],$$

the probabilkity being over all the random chocies made in the experiment.

Definition 2 We say that **TS** is transitive unforgeable under adaptive chosen-message attack if the function $\mathbf{Adv}_{TS,F}(k)$ is negligible for any adversary F whose running time is polynomial in the security parameter k.

2.   New undirected transitive signature scheme

In this section we describe an new transitive signature scheme for working on undirected graphs. it is based on the difficulty of the discrete logarithm problem.

Standard signature scheme

Our new scheme use an underling standard digital signature scheme **SDS**= (**SKG, SSign, SVf**), where **SKG** is polynomial time key generation, **SSign** is signing algorithm, and **SVf** is verification algorithm. We use the security definition proposed by Goldwasser, Micali and Rivest in [GoldMic].A forger B is given adaptive oracle access to the signing algorithm, and its advantage $\mathbf{Adv}_{SDS,B}(k)$ in breaking **SDS** is defined as the probability that it outputs a valid signature for a message that was not one of its previous oracle queries. The scheme **SDS** is said to be secure against forgery under adaptive chosen message attack if $\mathbf{Adv}_{SDS,B}(k)$ is negligible for every forgery B with running time polynomial in the security parameter k.

Discrete logarithm problem

A modulus generator is a randomized, polynomial time algorithm that on input 1^k returns a triple (p, q, g), where p and q are large primes, $2^{k-1} \le p < 2^k$, such that q divides $(p - 1)$ and $g \in Z_p^*$ is a generator of order, the group generated by g is denoted by G_q. We do not restrict the tpye of generator, but only assume that the associated discrete logarithm problem is hard. Formally, for any adverary A and any $k \in N$, we let

$\mathbf{Adv}_{(\mathrm{MG,A})^{\mathrm{DLP}}}(k)$
$= \Pr[y = g^x \bmod p : (p, q, g) \leftarrow \mathbf{MG}(1^k); \forall y \in Z_p^*; x \leftarrow A(k, p, q, g, y)]$

We say that discrete logarithm problem is hard if function $\mathbf{Adv}_{(\mathrm{MG,A})^{\mathrm{DLP}}}(k)$ is negigible for every A whose running time is polynomial in k.

New transitive signature scheme

Given a modulus generator and a standard signature scheme **SDS**=(**SKG, SSign, SVf**), we design a new transitive signature scheme **DLPTS**=(**TKG, TSign, TVf, Comp**) as follows.

- Given input 1^k, the key gerneration algorithm **TKG** first runs **SKG** on input 1^k to generate a key pair (spk, ssk) for the standard signature scheme **SDS**. It then runs the modulus generator **MG** on input 1^k to get a triple p, q, g. It outputs $tpk = (p, q, g, spk)$ as the public key. Let $S = \emptyset$, **Legit**=$true$, **NotOk**=$false$.

- The signing algorithm **TSign** maintains state (V, ℓ, Y, Σ), where $V \subseteq N$ is the set of all queried nodes, the function $\ell : V \to Z_N^*$ assigns to each node $i \in N$ a secret label $\ell(i) \in Z_N^*$, while the function $Y : V \to Z_N^*$ assigns to each node $i \in V$ a puvlic label y_i, and the function $\Sigma : V \to 0, 1^*$ assigns to each node i a standard signature on $(i||y_i)$ under secret key ssk. When invoked on inputs tsk, i, j, meaning asked to produce a signature on edge i, j, it does the following:

If $i = j$, then **Legit**$\leftarrow false$

If $j < i$ then $l \leftarrow j; j \leftarrow i; i \leftarrow l$ // swap i and j

If $i \notin V$ then $V \leftarrow V \cup \{i\}; \ell(i) \leftarrow Z_N^*; y_i = g^{\ell(i)} mod\ p;$

$\Sigma(i) \leftarrow \textbf{SSign}(ssk, (i\|y_i))$

If $j \notin V$ then $V \leftarrow V \cup \{j\}; \ell(j) \leftarrow Z_N^*; y_j = g^{\ell(j)} mod\ p;$

$\Sigma(j) \leftarrow \textbf{SSign}(ssk, (i\|y_j))$

$\delta \leftarrow [\ell(i) - \ell(j)] mod\ q; C_i \leftarrow (i, y_i, \Sigma(i)); C_j \leftarrow (i, y_j, \Sigma(j))$

$\sigma \leftarrow [C_i, C_j, \delta], S \leftarrow S \cup \{[(i, j), \sigma]\}$

If **TVf** $(tpk, i, j, \sigma) = 0$, then **NotOK** $\leftarrow true$

We refer to $(l, y_l, \Sigma(l))$ as a certificate of node l

Return (C_i, C_j, δ) as the signature of $\{i, j\}$.

Return **Legit** $\wedge$ **NotOK**

- The verification algorithm TVf, on input $tpk = (p, q, g, spk)$, nodes i, j and a candidate signature σ, proceeds as follows:

  If $j < i$ then $l \leftarrow j; j \leftarrow i; i \leftarrow l$ // swap i and j

  Let $\sigma = (C_i, C_j, \delta)$, and $C_i = (i, y_i, \Sigma(i))$, and $C_j = (j, y_j, \Sigma(j))$

  If $\textbf{SVf}(spk, i\|y_i, \Sigma(i)) = 0$ or $\textbf{SVf}(spk, j\|y_j, \Sigma(j)) = 0$,

  then return 0

  If $y_i(y_j)^{-1} = g^\delta mod\ p$ then return 1 else return 0.

- The composition algorithm **Comp** takes nodes i, j, k, a signature $\sigma_1 = (C_1, C_2, \delta_1)$ of the edge (i, j), and a signature $\sigma_2 = (C_3, C_4, \delta_2)$ of the edge (j, k), and processds as follows:

  If $[\{i, j\}) \notin S]$ or $[\{j, k\}) \notin S]$ or i, j, k are not all distinct,

  then **Legit** $\leftarrow false$

  Let $C_i \in \{C_1, C_2\}$, and $C_i = (i, y_i, \Sigma(i))$.

  Let $C_j \in \{C_1, C_2\}$, and $C_j = (j, y_j, \Sigma(j))$.

  If $C_j \notin \{C_3, C_4\}$, then return $\perp$.

  Let $C_k \in \{C_3, C_4\}$, and $C_k = (k, y_k, \Sigma(k))$.

  If $i < j < k$; then $\delta \leftarrow (\delta_1 + \delta_2)$; Return (C_i, C_k, δ)

  If $i < k < j$; then $\delta \leftarrow (\delta_1 - \delta_2)$; Return (C_i, C_k, δ)

  If $j < i < k$; then $\delta \leftarrow (-\delta_1 + \delta_2)$; Return (C_i, C_k, δ)

  If $k < i < j$; then $\delta \leftarrow (-\delta_1 + \delta_2)$; Return (C_i, C_k, δ)

  If $j < k < i$; then $\delta \leftarrow (\delta_1 - \delta_2)$; Return (C_i, C_k, δ)

If $k < j < i$; then $\delta \leftarrow (\delta_1 + \delta_2)$; Return (C_i, C_k, δ)

$\sigma \leftarrow [C_i, C_k, \delta]$, $S \leftarrow S \cup \{[(i, j), \sigma]\}$

Let $\tau \leftarrow \mathbf{Tsign}(tsk, i, k)$

If $(\sigma \neq \tau)$ or $\mathbf{TVf}(tpk, i, j, \sigma) = 0$, then $\mathbf{NotOK} \leftarrow true$

Return $\mathbf{Legit} \wedge \mathbf{NotOK}$

3.   Correctness

In this section, we will prove the correctness of this new scheme. We first have the following lemma.

Lemma 1: Suppose that $G^* = (V^*, E^*)$ is a transitively closed graph. We will make transitive signature on this graph with above **DLPTS**. Let S be the set of edges and corresponding signature in processing **Tsign** algorithm. For $\forall((i, j), \sigma) \in S$, we have the following equations.

$$\{i \neq j\} \wedge \sigma = \left\{ \begin{array}{l} [(i, Y(i), \Sigma(i)), (j, Y(j), \Sigma(j)), (\ell(i) - \ell(j))] \text{ if } i < j \\ [(j, Y(j), \Sigma(j)), (i, Y(i), \Sigma(i)), (\ell(j) - \ell(i))] \text{ if } i < j \end{array} \right. \quad (1)$$

Proof: In **DLPTS** transitive signature scheme, there are two algorithms generating new element $(\{i, j\}, \sigma)$ to be added to S, one is **Tsign**, other is **Comp**.

At the beginning of **DLPTs**, $S = \emptyset$

Firstly, we consider the **Tsign** oracle query with $\{i, j\}$.

If $i = j$, **legit** is set to $false$, and the stop **Tsign**, no new element is added to S, so the above claim is right.

Else, a new element $(\{i, j\}, \sigma)$ is added to S, where σ is the output of **Tsign**(tsk, i, j).

If $i < j, \sigma = [(i, Y(i), \Sigma(i)), (j, Y(j), \Sigma(j)), \delta)]$, where $\delta = (\ell(i) - \ell(j))$;
If $j < i, \sigma = [(j, Y(j), \Sigma(j)), (i, Y(i), \Sigma(i)), \delta)]$, where $\delta = (\ell(j) - \ell(i))$;

Therefor, the newly added element $(\{i, j\}, \sigma)$ satisfies the above Equation (1). Because **Tsign** only adds new element to S, but never changes existing elements in S, after the **Tsign** oracle query, all elements of S still satisfie the above Equation (1). The above claim is right.

Second, we onsider the **Comp** oracle query with $i, j, k, \sigma_1, \Sigma_2$.

If $(\{i, j\}, \sigma_1 \notin S)$ or $(\{j, k\}, \sigma_2 \notin S)$ or i, j, k are not all distinct, then **legit** is set to $false$, and the stop **Comp**, no new element is added to S, so the above claim holds true.

Else, the composition algorithm is run, and a new element $(\{i, k\}, \delta)$ is created, and is added to S, where $\delta = Comp(tpk, i, j, k, \sigma_1, \sigma_2)$.

Depending on the relations between i, j and k, the variable δ inside the compostion algorithm gets diferent values as following.

If $i < j < k$, $\delta_1 = (\ell(i) - \ell(j))$, $\delta_2 = (\ell(j) - \ell(k))$,
$\delta = (\delta_1 + \delta_2) = (\ell(i) - \ell(k))$;
If $i < k < j$, $\delta_1 = (\ell(i) - \ell(j))$, $\delta_2 = (\ell(k) - \ell(j))$,
$\delta = (\delta_1 - \delta_2) = (\ell(i) - \ell(k))$;
If $j < i < k$, $\delta_1 = (\ell(j) - \ell(i))$, $\delta_2 = (\ell(j) - \ell(k))$,
$\delta = (-\delta_1 + \delta_2) = (\ell(i) - \ell(k))$;
If $k < i < j$, $\delta_1 = (\ell(i) - \ell(j))$, $\delta_2 = (\ell(k) - \ell(j))$,
$\delta = (-\delta_1 + \delta_2) = (\ell(k) - \ell(i))$;
If $j < k < i$, $\delta_1 = (\ell(j) - \ell(i))$, $\delta_2 = (\ell(j) - \ell(k))$,
$\delta = (\delta_1 - \delta_2) = (\ell(k) - \ell(i))$;
If $k < j < i$, $\delta_1 = (\ell(j) - \ell(i))$, $\delta_2 = (\ell(k) - \ell(j))$,
$\delta = (\delta_1 + \delta_2) = (\ell(k) - \ell(i))$;

Therefor, the newly added element $(\{i, k\}, \delta)$ satisfies the above Equation (1). Because **Comp** only adds new element to S, but never changes existing elements in S, so all elements of S still satisfie the above Equation (1). The above claim is right. Lemma 1 has been proved.

In order to verfify the validity of any $((i, j), \sigma) \in S$ in the **DLPTS** transitive signature scheme, where $\sigma = [(i, Y(i), \Sigma(i)), (j, Y(j), \Sigma(j)), \delta)]$, we must verify the following two equations.

(1) **SVf**$(spk, i||Y(i), \Sigma(i))$ and **SVf**$(spk, j||Y(j), \Sigma(j))$, and
(2) $Y(i)Y(j)^{-1} = g^\delta mod\, p$

For $\forall i \in V$, $\Sigma(i)$=**SSign**$(ssk, i||Y(i))$, **SSign** is one section of a standard signature scheme **SDS**, **SVf** is the signature verifing sectin in the same **SDS**, so the **SVf**$(spk, i||Y(i), \Sigma(i))$ is true.

On the other hand, from the Lemma 1, $\forall((i, j), \sigma) \in S, \sigma = [(i, Y(i), \Sigma(i)), (j, Y(j), \Sigma(j)), \delta)]$, we have the equation $Y(i) = g^\delta Y(j) \mod p$.

Lemma 2: At any time in **DLPTS** transitive signature scheme, $\forall((i, j), \sigma) \in S$, we have the following equations.

$$\mathbf{TVf}(tpk, i, j, \sigma) \equiv 1 \qquad (2)$$

Lemma 3: The variable **NotOK** in **DLPTS** transitive signature scheme can never become $true$.

Proof: By the above lemmas, the verification of a signature in S always successds, meaning that for any $((i, j), \sigma) \in S$, we have **TVf**$(tpk, i, j, \sigma) \equiv 1$, so the only way left for **NotOK** to become $true$ in **DLPTS** transitive signature scheme is when $\sigma \neq \tau$ in a compostion algorithm $\mathbf{Comp}(i, j, k, \sigma_1, \sigma_2)$. The output of the signature algorithm $\mathbf{Tsign}((i, k), \sigma)$ for edge (i, k) is τ as follows.

$$\tau = \{ \begin{array}{l} [(i, Y(i), \Sigma(i)), (k, Y(k), \Sigma(k)), (\ell(i) - \ell(k))] \text{ if } i < k \\ [(k, Y(k), \Sigma(k)), (i, Y(i), \Sigma(i)), (\ell(k) - \ell(i))] \text{ if } k < i \end{array} \tag{3}$$

The output of composition algorithm $\mathbf{Comp}(i, j, k, \sigma_1, \sigma_2)$ in $\mathbf{DLPTS}$ transitive signature scheme is σ as follows.

$$\sigma = [(i, Y(i), \Sigma(i)), (k, Y(k), \Sigma(k)), \delta] \tag{4}$$

where $\delta = \{ \begin{array}{l} (\ell(i) - \ell(k)) \text{ if } i < k \\ (\ell(k) - \ell(i)) \text{ if } k < i \end{array}$. So we can see that τ is exactly σ. The Lemma 3 has been proved

Theorem 1: The $\mathbf{DLPTS}$ transitive signature scheme described in section 3 staisfies the correctness requirment of Definitin 1.
Proof: The $\mathbf{DLPTS}$ transitive signature scheme will return $(\mathbf{Legit} \wedge \mathbf{NotOK})$ at the end of its execution. From the above lemmas, we see that it is impossible for $(\mathbf{Legit} \wedge \mathbf{NotOK})$ to return $true$. Thereby, we have proved the correctness of $\mathbf{DLPTS}$ transitive signature scheme.

4.   Security

This section discusses the security of the new scheme under the assumption of intractability of discrete logarithm problem. As with any signature scheme, proving security means proving an adversary will not able to forge new signature even if having seen some previous legitimate signatures.

In the transitive signature scheme $\mathbf{DLPTS} = (\mathbf{TKG, TSign, TVf, Comp})$, $\mathbf{TVf}$ is the verification algorithm. On input a signature $\sigma = (C_i, C_j, \delta)$ of $\{i, j\}$, where $C_i = (i, Y(i), \Sigma(i))$, $C_j = (i, Y(j), \Sigma(j))$, the verification alogarith verifies the following equations.

$$\mathbf{SVF}(spk, t||Y(t), \Sigma(t) = 1, \text{ where } (t = i, j). \tag{5}$$

$$Y(i)Y(j)^{-1} = g^\delta mod\ p, \text{ if } i < j, \text{ else swap } i \text{ and } j \tag{6}$$

The former equation is about the signature verification of the digital standard signature $\mathbf{SDS} = (\mathbf{SKG, SSign, SVF})$, the later equation is additional verification of transitive signature scheme $\mathbf{DLPTS}$. Having passed all above verifications, we say that this signature is valid.

Given parameter (p, q, g), we have the following lemma.

Lemma 4: For any $(Y(i), Y(j))$, solving the $\delta \in G_q$ form following equation is equal to discrete logarithm problem, where $(Y(i), Y(j))$ are random in Z_p^*.

$$Y(i)Y(j)^{-1} = g^\delta mod\ p$$

Proof: We denote the above problem to $\mathbf{DLP^*}$.

First, we have $\mathbf{DLP} => \mathbf{DLP^*}$.

Let $x_i = \mathbf{DLP}(Y(i))$, $x_j = \mathbf{DLP}(Y(j))$, then $\delta = (x_i - x_j) mod\ q$, which satifies $Y(i)Y(j)^{-1} = g^\delta mod\ p$.

Second, we have $\mathbf{DLP} <= \mathbf{DLP^*}$.

Let $Y = Y(i)Y(j)^{-1} mod\ p$, $\delta = \mathbf{DLP^*}(Y(i), Y(j))$, and $x = \delta$, then $Y = g^x mod\ p$. Because $Y(i)$ and $Y(j)$ are random in Z_n^*, so Y also are random. This means that for a $Y \in Z_n^*$, which is random, we can get a x satisfing $Y = g^x mod\ p$.

Lemmma 4 has been proved.

Suppose we are given a polynomial-time algorithm forger F for $\mathbf{DLPTS}$. A is a algorithm to compute $\mathbf{DLP}$ whose parameters are generated by $\mathbf{MG}$, and B is attacker to the standard digital signature $\mathbf{SDS}$. $tpk = (p, q, g)$, spk is a public key for the transitive signature scheme. For $\forall (i, j) \in N$, once F is done querying its oracle, it will output a signature $((i, j), \sigma)$ of the edge $\{i, j\}$, where $\sigma = ((i, Y_i, \Sigma_i)(j, Y_j, \Sigma_j), \delta)$.

Let $G^* = (V, E^*)$ denote the transitive closure of $G = (V, E)$, and B_1, B_2, B_3 are three objects, which are difined as following.

$B_1 = \{\mathbf{TVf}(tpk, i, j, \sigma) \neq 1\}$
$B_2 = \{(i, j) \notin E^*\}$
$B_3 = \{Y_i \neq Y(i)\ or\ Y_j \neq Y(j)\}$

Because (i, j) are randomly chose in Z, and $\ell : V \to Z_q^*$ is a random function, so $Y(i), Y(j)$ are arbitrary in Z_p^*. From lemma 4, we have the following equation.

$$\mathbf{Adv}_{MG,A}^{DLP}(k) = Pr(\overline{B_1}) \geq Pr(\overline{B_1} \wedge \overline{B_2} \wedge \overline{B_3}) \tag{7}$$

$$= Pr(\overline{B_3}|\overline{B_1} \wedge \overline{B_2})Pr(\overline{B_1} \wedge \overline{B_2}) \tag{8}$$

$$= Pr(\overline{B_3}|\overline{B_1} \wedge \overline{B_2})\mathbf{Adv}_{DLPTS,F}(k) \tag{9}$$

Algorithm B will perform a chosen-message attack on standard digital signature $\mathbf{SDS}$ using F as a subroutine. It is given access to a signing oracle $\mathbf{SSign}_s sk(.)$ and is considered successful if at the end of its execution, it outputs a valid signature pair relative to spk where the message was not one of its former oracle queries. If F outputs its forgery $((i, j), \sigma)$ of the edge $\{i, j\}$, where $\sigma = ((i, Y_i, \Sigma_i)(j, Y_j, \Sigma_j), \delta)$, suppose that B_1 and B_2 are not true, and $Y_i = Y(i)$ and $Y_i = Y(j)$, at least one of the signatures Σ_i and Σ_j must be a forgery. B is successful when these evens $\overline{B_1}, \overline{B_2}$, and B_3 happan simultaneously.

we have the following equation.

$$\mathbf{Adv}_{SDS,B}(k) = Pr(\overline{B_1} \wedge \overline{B_2} \wedge B_3) \tag{10}$$

$$= Pr(B_3|\overline{B_1} \wedge \overline{B_2})Pr(\overline{B_1} \wedge \overline{B_2}) \tag{11}$$

$$= (1 - Pr[\overline{B_3}|\overline{B_1} \wedge \overline{B_2}])\mathbf{Adv}_{DLPTS,F}(k) \tag{12}$$

From formulas (9) and (12),we have Lemma 5.

Lemma 5: In **DLPTS** transitive signature scheme, F is a polynomial-time algorithm forger, A is a algorithm to compute **DLP** whose parameters are generated by **MG**, and B is attacker to the standard digital signature **SDS**. For all k, we have the follwing equation.

$$\mathbf{Adv}_{\mathrm{DLPTS,F}}(k) \leq \mathbf{Adv}_{\mathrm{MG,A}}^{\mathrm{DLP}}(k) + \mathbf{Adv}_{\mathrm{SDS,B}}(k) \tag{13}$$

The hardness assumptions of **DLP** and **SDS** mean that both $\mathbf{Adv}_{\mathrm{MG,A}}^{\mathrm{DLP}}(k)$ and $\mathbf{Adv}_{\mathrm{SDS,B}}(k)$ are negligible in k for all alogarithma A and B with running time polynomial in k. By the above lemma 5, this implies that $\mathbf{Adv}_{\mathrm{DLPTS,F}}(k)$ is negligible for any polynomial time forger F.

We have the following theorm 2 with **DLPTS** transitive signature scheme.

Theorm 2: If the **DLP** is hard, and standard digital signature **SDS** is secure against forgery under adaptive chosen message attack, then **DLPTS** transitive signature scheme is unforgerable under adaptive chosen message attack.

We have proved the unforgeability of **DLPTS** transitive signature scheme.

5.   Conclusion

In this paper, according to the formal definition of transitive signature, described by Bellare and Neven in [BellNev], we design an new transitive signature scheme based on the discrete logarithm problem. The new scheme is different with Micali and Rivest's scheme. In the our scheme, there are two signature algorithms , one is for signing node, another is for signing edge. So it is more generlzed tranistive signature scheme based on discrete logarithm problem. With random oracle model, we also prove the correctness and the tansitive unforgeable of the new scheme under adaptive chosen-message attack assuming discrete logarithm problem is hard.

References

[MicaRiv] S. Micali, R. Rivest, *Transitive signature schemes*, Topics in Cryptology - CT-RSA'02, Lecture Notes in Computer Scinece Vol. 2271, B. Preneel ed., Springer-Verlag, 2002.

[JohnMol] R. Johnson, D. Molnar, D. Song, D. Wagner, *Homomorphic signature schemes*, Topics in Cryptology - CT-RSA'02, Lecture Notes in Computer Scinece Vol. 2271, B. Preneel ed., Springer-Verlag, 2002.

[BellNev] M. Bellare, G. Neven, *Transitive signatures based on factoring and RSA, Advances in Cryptology*, Aisacypto'02, Leture Notes in Computer Science Vol.2501, 2002,

[GoldMic] S. Goldwasser, S. Micali, R. Rivest, *A digital signature scheme scure against adaptive chosen-message attacks, SIAM Journal of Computing*, Vol.17, No.2, April 1988

BLIND SIGNATURE SCHEMES BASED ON GOST SIGNATURE *

Zhenjie Huang, Yumin Wang

National Key Lab of ISN, Xidian University, Xi'an, Shaanxi, 710071, P. R. China

zhj_huang@hotmail.com, ymwang@xidian.edu.cn

Abstract   In this paper, the authors propose a generalized blind GOST signature scheme and three practical blind GOST signature schemes educed from the proposed generalized scheme by setting one of three parameters be a constant.

Keywords: cryptography, digital signature, blind signature, GOST

Introduction

Blind signature, first introduced by D.Chaum in [1] at Ctypto'82, is a variant of digital signature schemes, which allows the requester to get a signature without giving the signer any information about the actual message or the resulting signature. Several signature schemes have been turned into blind signature schemes.

The first blind signature scheme that based on RSA signature scheme was proposed by D.Chaum in [1]. In [2] T.Okamoto proposed the blind Schnorr signature and D. Pointcheval proved its security in [3]. The blinding schemes of the modification of DSA and Nyberg-Rueppel message recovery signature scheme are presented in [4] by J.L.Camenisch. D. Pointcheval gave a blinding of Okamoto signature in [5] and proved its security in [3]. E.Mohammed proposed a blind signature scheme based on ElGamal signature in [6], unfortunately, this scheme is insecure (the signer's secret key can be derived). In [7], M.Abe presented a blind signature scheme that needs only three data moves and provides polynomial security.

The GOST signature scheme is the Russia's digital signature algorithm [8, 9]. In this paper, we propose a generalized blind GOST signature scheme and three practical blind GOST signature schemes educed from the proposed generalized scheme by setting one of three parameters be a constant.

* This work is supported by the Natural Science Foundation of China (Grant No. 19931010 and 60073052).

1.   GOST signature scheme

Let p, q be large primes that satisfy $q|p-1$, and g be an element in $\mathbf{Z}_p^*$ with order q. Let $H : \{0,1\}^* \to \mathbf{Z}_q$ be a secure hash function. The signer's public and secret key pair is (y, x), where $x \in \mathbf{Z}_q, y = g^x \bmod p$. Let m be the message to be signed.

- **Signing**: The signer chooses random number $k \in_R \mathbf{Z}_q$, and computes

$$r = (g^k \bmod p) \bmod q$$
$$s = xr + kH(m) \bmod q$$

(r, s) is the signature on message m.

- **Verifying**: The verifier computes

$$v = H(m)^{q-2} \bmod q$$
$$z_1 = sv \bmod q$$
$$z_2 = (q - r)v \bmod q$$
$$u = (g^{z_1} y^{z_2} \bmod p) \bmod q$$

and checks whether $u = r$.

2.   Blind GOST signature schemes

In this section, we first propose a generalized blind GOST signature scheme, and then give three practical blind GOST signature schemes educed from the generalized scheme by setting one of three parameters be a constant.

2.1   Generalized scheme

2.1.1   The scheme.   The cryptographic setting is as above. Following is the generalized blind GOST signature scheme.

- **Signing**: 1. The signer chooses random number $k \in_R \mathbf{Z}_q$, and computes

$$r = (g^k \bmod p) \bmod q$$

then sends r to the requester.

2. The requester chooses random numbers $t_1, t_2, t_3 \in_R \mathbf{Z}_q$, and computes

$$R = (r^{t_1} g^{t_2} y^{t_3} \bmod p) \bmod q$$
$$E = H(m)$$
$$e = rt_1(RE^{-1} + t_3)^{-1} \bmod q \qquad (1)$$

then sends e to the signer.

3. The signer calculates

$$s = xr + ke \bmod q$$

and then sends s to the requester.

4. The requester checks whether

$$g^s = y^r r^e \bmod p$$

If the requester accepts, computes

$$S = E(se^{-1}t_1 + t_2) \bmod q \qquad (2)$$

and publishes (R, E, S) as the signature on message m.

- **Verifying**: The verifier computes

$$v = E^{q-2} \bmod q$$
$$z_1 = Sv \bmod q$$
$$z_2 = (q - R)v \bmod q$$
$$u = (g^{z_1} y^{z_2} \bmod p) \bmod q$$

and checks whether $u = R$ and $E = H(m)$.

2.1.2   The security.   Here we discuss the security of the generalized scheme above.

Completeness.   The completeness can easily be proved as follows.

From equations $(1), (2)$, we obtain

$$-RE^{-1} = t_3 - re^{-1}t_1 \bmod q$$
$$SE^{-1} = se^{-1}t_1 + t_2 \bmod q$$

and then have that

$$\begin{aligned} g^{z_1} y^{z_2} &= g^{SE^{q-2}} y^{(q-R)E^{q-2}} \\ &= g^{SE^{-1}} y^{-RE^{-1}} \\ &= g^{se^{-1}t_1 + t_2} y^{t_3 - re^{-1}t_1} \\ &= g^{(s-xr)e^{-1}t_1} g^{t_2} y^{t_3} \\ &= g^{kt_1} g^{t_2} y^{t_3} \\ &= r^{t_1} g^{t_2} y^{t_3} = R \bmod p \bmod q \end{aligned}$$

where $s = xr + ke \bmod q$.

Unforgeability.   Since the verifying equation of our scheme is the same as that of the GOST scheme, and the blind signature (R, E, S) can be see as a signature of the GOST scheme, then, if an adversary can forge a valid blind signature, he also can forge a valid GOST signature. On the other hand, if an adversary can forge a valid GOST signature, he also can obtain a valid blind signature. Thus, the unforgeability of our scheme is the same as that of the GOST scheme.

Blindness. A signature scheme is called blind if the signer's view and the resulting signature are statistically independent, where the signer's view is the set of all values that can be gotten by the signer during the execution of the signature issuing protocol.

Since there are three random parameters in the three blinding functions

$$R = (r^{t_1} g^{t_2} y^{t_3} \bmod p) \bmod q$$
$$e = rt_1(RE^{-1} + t_3)^{-1} \bmod q$$
$$S = E(se^{-1}t_1 + t_2) \bmod q$$

one can easily see that there always exists a tuple of random factors (t_1, t_2, t_3) that maps (r, e, s) to (R, E, S) for any (r, e, s) and any (R, E, S). So the scheme is blind.

2.2   Educed schemes

In fact, two random parameters are sufficient to provide blindness. So (t_1, t_2, t_3) can be one of $(1, t_2, t_3), (t_1, 0, t_3)$ and $(t_1, t_2, 0)$, thus we can obtain three schemes from the generalized scheme above as following.

Case 1. $t_1 = 1$
In this case, the blinding functions as follows

$$R = (rg^{t_2} y^{t_3} \bmod p) \bmod q$$
$$e = r(RE^{-1} + t_3)^{-1} \bmod q$$
$$S = E(se^{-1} + t_2) \bmod q$$

The completeness and the unforgeability are the same as the generalized scheme, and the blindness can be proved as follows.

For $i = 0, 1$, let r_i, e_i, s_i, k_i be data appearing in the view of the signer during the execution of the signature issuing protocol on message m_i, and let R_i, E_i, S_i be the corresponding signatures. It is sufficient to show that there exists a tuple of random factors (t_2, t_3) that maps r_i, e_i, s_i, k_i to R_j, E_j, S_j for each $i, j \in \{0, 1\}$. To this end, we define

$$t_2 = S_j E_j^{-1} - s_i e_i^{-1} \bmod q$$
$$t_3 = r_i e_i^{-1} - R_j E_j^{-1} \bmod q$$

and have that

$$
\begin{aligned}
r_i g^{t_2} y^{t_3} &= g^{k_i} g^{S_j E_j^{-1} - s_i e_i^{-1}} y^{r_i e_i^{-1} - R_j E_j^{-1}} \\
&= g^{k_i - s_i e_i^{-1} + x r_i e_i^{-1}} g^{S_j E_j^{-1}} y^{-R_j E_j^{-1}} \\
&= g^{S_j E_j^{-1}} y^{-R_j E_j^{-1}} \\
&= g^{S_j E_j^{q-2}} y^{-R_j E_j^{q-2}} = R_j \bmod p \bmod q
\end{aligned}
$$

where $s_i = xr_i + k_i e_i \bmod q$.

Thus, r_i, e_i, s_i, k_i and R_j, E_j, S_j have exactly the same relation defined by the signature issuing protocol, and the proposed scheme is blind.

Case 2. $t_2 = 0$

In this case, the blinding functions as follows

$$R = (r^{t_1} y^{t_3} \bmod p) \bmod q$$

$$e = rt_1 (RE^{-1} + t_3)^{-1} \bmod q$$

$$S = Ese^{-1} t_1 \bmod q$$

The completeness and the unforgeability are also the same as the generalized scheme, the blindness is similar to the case 1.

We define

$$t_1 = S_j E_j^{-1} s_i^{-1} e_i \bmod q$$

$$t_3 = S_j E_j^{-1} s_i^{-1} r_i - R_j E_j^{-1} \bmod q$$

and have that

$$
\begin{aligned}
r_i^{t_1} y^{t_3} &= g^{k_i S_j E_j^{-1} s_i^{-1} e_i} y^{S_j E_j^{-1} s_i^{-1} r_i - R_j E_j^{-1}} \\
&= g^{S_j E_j^{-1} s_i^{-1} (e_i k_i + x r_i)} y^{-R_j E_j^{-1}} \\
&= g^{S_j E_j^{-1}} y^{-R_j E_j^{-1}} \\
&= g^{S_j E_j^{q-2}} y^{-R_j E_j^{q-2}} \\
&= R_j \bmod p \bmod q
\end{aligned}
$$

where $s_i = xr_i + k_i e_i \bmod q$.

Thus, r_i, e_i, s_i, k_i and R_j, E_j, S_j have exactly the same relation defined by the signature issuing protocol, and the proposed scheme is blind.

Case 3. $t_3 = 0$

In this case, the blinding functions as follows

$$R = (r^{t_1} g^{t_2} \bmod p) \bmod q$$

$$e = rt_1 R^{-1} E \bmod q$$

$$S = E(se^{-1} t_1 + t_2) \bmod q$$

The completeness and the unforgeability are also the same as the generalized scheme, the blindness is similar to the case 1 and case 2.

We define

$$t_1 = R_j E_j^{-1} r_i^{-1} e_i \bmod q$$

$$t_2 = S_j E_j^{-1} - R_j E_j^{-1} r_i^{-1} s_i \bmod q$$

and have that

$$
\begin{aligned}
r_i^{t_1} g^{t_3} &= g^{k_i R_j E_j^{-1} r_i^{-1} e_i} g^{S_j E_j^{-1} - R_j E_j^{-1} r_i^{-1} s_i} \\
&= g^{R_j E_j^{-1} r_i^{-1}(e_i k_i - s_i)} g^{S_j E_j^{-1}} \\
&= g^{-x R_j E_j^{-1}} g^{S_j E_j^{-1}} \\
&= g^{S_j E_j^{q-2}} y^{-R_j E_j^{q-2}} \\
&= R_j \bmod p \bmod q
\end{aligned}
$$

where $s_i = x r_i + k_i e_i \bmod q$.

Thus, r_i, e_i, s_i, k_i and R_j, E_j, S_j have exactly the same relation defined by the signature issuing protocol, and the proposed scheme is blind.

3.   Conclusion

In this paper, we propose a generalized blind GOST signature scheme and three practical blind GOST signature schemes.

References

[1] Chaum D. Blind Signatures for Untraceable Payments, Crypto'82, Prenum, 1982.

[2] Okamoto T. Provably Secure and Practical Identification Schemes and Corresponding Signature Schemes, Crypto'92, LNCS 740, 1992.

[3] Pointcheval D., Stern J. Security Arguments for Digital Signatures and Blind Signatures, J of Cryptology, 2000, 13(3).

[4] Camenisch J.L., Piveteau J.M., and Stadler M. A. Blind Signatures Based on the Discrete Logarithm Problem, Eurocrypt'94, LNCS 950,1994.

[5] Pointcheval D. Strengthened Security for Blind Signatures, Eurocrypt'98, LNCS 1403, 1998.

[6] Mohammed E., Emarah A.E., Shennawy K.E. A Blind Signature Scheme Based on ElGamal Signature, 17th NRSC'2000.

[7] Abe M. A Secure Three-move Blind Signature Scheme for Polynomially Many Signatures, Eurocrypt 2001, LNCS 2045, 2001.

[8] Michels M., Naccache D., Peterson H. GOST 34.10-a Brief Overview of Russia's DSA, Computer & Security, 1996,15(8).

[9] GOST R 34.10-94, Gosudarstvennyi Standard of Russian Federation, "Information technology. Cryptographic Data Security. Produce and check procedures of Electronic Digital Signature Based on Asymmetric Cryptographic Algorithm." Government Committee of the Russia for Standards, 1994.

ONE-OFF BLIND PUBLIC KEY

Zhang Qiupu, Guo Baoan

Dept of Computer Science and Technology Tsinghua University Beijing 100084 China
guoba@tsinghua.edu.cn

Abstract   The definition and properties of one-off blind public key are proposed. It just need the trusted entity issues the generative factor of blind public keys one time for the user, and the user can generate his different public keys each time he uses it. So this scheme can ensure the disconnection among the user's actions, and the trusted entity can reveal the user's identity under the court's grant to prevent the user committing, but anyone else doesn't have this ability. At the same time, this scheme can ensure the user can't forge one-off blind public key without the trusted entity. A new scheme of one-off blind public key is proposed.

Keywords: blind signatures, group signatures, Fiat-Shamir identification scheme, one-off blind public key

Introduction

The concept of one-off blind public key and the first scheme of one-off public key protocol are proposed in [6]. It just need the trusted entity issues the generative factor of blind public key one time for the user, and the user can generate his different public keys each time he uses it. So this scheme can ensure the disconnection among the user's actions, and the trusted entity can reveal the user's identity under the court's grant to prevent the user committing, but anyone else doesn't have this ability. But the definition and properties of one-off blind public key haven't presented.

1.   Definition and properties of one-off blind public key

Definition 1.1. *If based on the secret data x, the different public keys can be generated by using the public algorithm F. Those public keys can correspond with the same or the different private keys. Without knowing the secret data x, no one can calculate the secret data x, the private key and other public keys with the public algorithm F and the known public keys. So we name those public keys as blind public keys, the secret data x as the generative factor of blind public keys, the public algorithm F as blind public key generative algorithm.*

The trusted entity is used in one-off blind public key. One-off blind public key has essential properties as follows:

(1) one-off ability: It just need the trusted entity issues the generative factor of blind public key one time for the user, and the user can use his different public keys only once.

(2) public key changeability: Based on the generative factor of blind public key, the user can generate different public keys. Those public keys can correspond with the same or the different private keys.

(3) disconnection among the user's actions(includes anonymity):No one except the trusted entity can get any useful information about the user and his private key from the user's public keys. The receiver can know the sender has the legal public key and private key, but can't discover the sender's identity. If the public key the user used is different each time, the receiver can't get the connection from the user's actions and the user's blind public keys.

(4) unforgeability: No one can forge one-off blind public key without the trusted entity. The one-off blind public key that the user generates must be generated from the generative factor of blind public key that was issued for the user.

(5) unfraudulence: When the disputation occurs, the one-off blind public key can be send to the trusted entity, the trusted entity can discover the user's identity, so it can prevent the user's committing with one-off blind public key.

2.   Relative knowledge

2.1   The theorem comes from [5]

Let $n = p \cdot q$, where $p < q$, $p = 2p' + 1$, $q = 2q' + 1$,and p, q, p', q' are all prime numbers. Then,

(1) The order of elements in Z_n^* is one of the number in the set $\{1, 2, p', q', 2p', 2q', p'q', 2p'q'\}$.

(2) Given a elements $w \in Z_n^* \backslash \{-1, 1\}$, such that $ord(w) < p' \cdot q'$ then either $gcd(w - 1, n)$ or $gcd(w + 1, n)$ is a prime factor of n.

2.2   The Fiat-Shamir identification scheme

The Fiat-Shamir identification scheme is an efficient method enabling one party to authenticate its identity to another party. A modification of the Fiat-Shamir identification scheme is used in our scheme [1]:

The trusted entity T picks two prime numbers p, q, $n = p \cdot q$, T picks randomly a invertible elements $x_A \mod n$, and x_A has a large order. T calculates $y_A = x_A^k$, and k is a prime, $k < ord(x_A)$, so A's secret key is x_A, A's public key is (y_A, k). The following scheme can ensure A has x_A, and the public key (y_A, k) he used satisfies $y_A = x_A^k \mod n$.

(1) A picks a random $r, r < n$, sends $r^k \mod n$ to B.

(2) B picks a random bit b, sends it to A.

(3) A calculates $c = r \cdot x_A^b \mod n$, sends c to B.

(4) B verifies $c^k = r^k \cdot y_A^b \mod n$.

The possibility of this protocol's fraudulence is 1/2. To ensure security, we need run this protocol t times repeatedly, the random r is differently each time, so the possibility of fraud is $1/2^t$.

2.3 Group signature [2]

Group signature allows one to sign on behalf of the group. Normally, the scheme of group signature is composed of group, the member of the group (the signer), the receiver (the message verifier) and the trusted entity. The properties of group signature are as follows:

(1) Only the member of the group can sign, the signature is on behalf of the group.

(2) The receiver can verify the validity of the signature, but can't discover the sender's identity.

(3) When the disputation occurs, the trusted entity can discover the sender's identity.

3. One-off blind public key protocol

3.1 The initialization of the trusted entity

(1) The trusted entity T selects the prime numbers $p, q, p', q', p = 2p' + 1, q = 2q' + 1$, the length of p', q' is equal, $n = pq, \varphi(n) = (p-1)(q-1)$.

(2) The trusted entity T selects a random $e, gcd(e, \varphi(n)) = 1$. T calculates $d, ed \equiv 1 \mod \varphi(n)$. n, e is the trusted entity's public key, d is the trusted entity's private key. T publicize $len(p'q')$, $len(p'q')$ is the length of $p'q'$.

(3) Let $G = \{g | ord(g) = p'q', g \in Z_n^*\}$.

3.2 Issue generative factor of blind public key for user

(1) User A uses his ID to register in the trusted entity T. T picks $e_A \in G$ as A's generative factor of blind public key, and ensures every member's generative factor of blind public key is different. T records A's ID and e_A to its database.

(2) T signs on e_A based on RSA, viz calculates $v_A = e_A^d \mod n$, sends secretly (e_A, v_A) to A.

3.3 Calculation of blind public key

(1) After getting (e_A, v_A), A verifies T's signature.

(2) A picks prime number k, $k > 2$, $len(k) < len(p'q')$, A calculates $e_{A,k} = e_A^k \mod n$. A picks two prime numbers $p_{A,k}$, and $q_{A,k}$, $gcd(e_{A,k}, (p_{A,k} - 1) \cdot (q_{A,k} - 1)) = 1$, $n_{A,k} = p_{A,k} \cdot q_{A,k}$. A calculates $d_{A,k}$, $e_{A,k} \cdot d_{A,k} = 1 \mod (p_{A,k} - 1)(q_{A,k} - 1$. A calculates $v_{A,k} = v_A^k = (e_A^d)^k = e_{A,k}^d \mod n$. Now, $d_{A,k}$ is A's private key, $n_{A,k}, e_{A,k}$, are corresponding public key, $v_{A,k}$ is T's blind signature on $e_{A,k}$, then $(n_{A,k}, e_{A,k}, v_{A,k}, k)$ is A's one-off blind public key. A records k, $p_{A,k}, q_{A,k}$ to his database. When calculating his one-off blind public key next time, A ensures k, $p_{A,k}$, $q_{A,k}$ is distinctive.

In this scheme, e_A is the generative factor of blind public key, so e_A must be secret.

3.4   Verification of the validity of one-off blind public key.

(1) A picks a random r, $r < n$, calculates $u = r^k \mod n$, sends $(n_{A,k}, e_{A,k}, v_{A,k}, k, u)$ to B after encrypting with B's public key.

(2) B picks a random bit b, sends it to A.

(3) A calculates $c = r \cdot e_A^b \mod n$, sends c to B.

(4) B makes sure that A has e_A and $e_{A,k} = e_A^k \mod n$ by verifying $c^k = r^k e_{A,k}^b \mod n$. The proof of this verification is $c^k = (r \cdot e_A^b)^k = r^k \cdot e_A^{bk} = u \cdot e_{A,k}^b \mod n$.

(5) B verifies k is prime number, $k > 2$, and $len(k) < len(p'q')$, it can ensure that the trusted entity can reveal the user's identity under the court's grant to prevent the user committing.

(6) B makes sure that A has registered in the trusted entity, and the one-off blind public key that A generated bases on the generative factor of blind public key that the trusted entity issues for A. The proof of this verification is $v_{A,k}^e = (e_{A,k}^d)^e = e_{A,k} \mod n$.

(1)-(4) is the modification of Fiat-Shamir identification scheme. In this step, the possibility of fraudulence is 1/2. To ensure security, it need run (1)-(4) t times repeatedly, and the random r is differently each time. Thus can reduce the possibility of fraud to $1/2^t$.

3.5   Useing of one-off blind public key and the private key

$(n_{A,k}, e_{A,k}, d_{A,k})$ can be used in RSA's encryption scheme and signature scheme. $(n_{A,k}, e_{A,k})$ is public key, $d_{A,k}$ is private key. When $(n_{A,k}, e_{A,k}, d_{A,k})$ is used for encrypting, it is the session key based on public cryptography, and it has the trusted entity's warrant.

(1) A generates his one-off blind public key, sends it to B.

(2) B encrypts message with this one-off blind public key in this communication action.

(3) A decrypts message with the corresponding private key.

(4) In next communication action, A generates and uses the new one-off blind public key.

4.   Security analysis of one-off blind public key

4.1   One-off ability

It just need the trusted entity issues the generative factor of blind public key one time for A, and A can generate one-off blind public key use his generative factor of blind public key. Among his different communication actions, A can't use the same one-off blind public key, otherwise his actions can be connected with the same one-off blind public key.

4.2   Public key changeability

Based on the generative factor of blind public key, the user can generate different public keys. Every public key has the distinctive private key.

4.3   Disconnection among the user's actions

In the whole process, B just know A has the generative factor of blind public key e_A issued by the trusted entity, but can't get any other information about A, and can't get A's fixed public key. If A doesn't use the same k, $p_{A,k}$ and $q_{A,k}$ every time, his $(n_{A,k}, e_{A,k}, v_{A,k}, k)$ is different, his public key is one-off blind public key, and his actions can't be connected with his one-off blind public key.

The trusted entity can't know A's identity by view, because A's one-off blind public key is encrypted with B's public key in transmission. Only when the one-off blind public key is committed to the trusted entity under the court's grant, the trusted entity can calculate A's generative factor of blind public key e_A with $(n_{A,k}, e_{A,k}, v_{A,k}, k)$.

4.4   Unforgeability

If A has the legal generative factor of blind public key e_A, but $e_{A,k}$ and k in $(n_{A,k}, e_{A,k}, v_{A,k}, k)$ doesn't satisfy $e_{A,k} = e_A^k \mod n$, A will be exposed when using the Fiat-Shamir identification scheme.

C can't picks a number $v_{C,k}$, calculates $e_{C,k} = v_{C,k}^e \mod n$, picks a prime number $k > 2$ and $n_{C,k}$, declares $(n_{C,k}, e_{C,k}, v_{C,k}, k)$ as his one-off blind public key. Because C doesn't know $\varphi(n)$, he can't calculate $t = k^{-1} \mod \varphi(n)$, and can't calculate $e_C = e_{C,k}^t \mod n$. It means C hasn't registered in the trusted entity. A will be exposed when using the Fiat-Shamir identification scheme.

C can't forge $(n_{C,k}, d_{C,k}, e_{C,k})$ without the trusted entity, and use it in encryption and signature. Because C doesn't know $\varphi(n)$, he can't calculate $d = e^{-1}$

mod $\varphi(n)$, and can't calculate $v_{C,k} = e_{C,k}^d$ mod n. It means C can't forge $v_{C,k}$. If C picks a random $v_{C,k}$, it can't pass the verification $v_{C,k}^e = e_{C,k}$ mod n.

Any one including B can't make use of the one-off blind public key $(n_{A,k}, e_{A,k}, v_{A,k}, k)$ that was made by A, then picks number $r(r < n)$, calculates $e_{A,k'} = e_{A,k}^r$ mod n, $v_{A,k'} = v_{A,k}^r$ mod n to forge one-off blind public key. Now, $e_{A,k'} = e_{A,k}^r = (e_A^k)^r = e_A^{kr} = e_A^{k'}$ mod n, $v_{A,k'} = v_{A,k}^r = (v_A^k)^r = v_A^{kr} = v_A^{k'}$ mod n. If hoping pass the Fiat-Shamir identification scheme, it needs $k' = kr$ mod $\varphi(n)$, and needs k' is a prime number, but no one knows $\varphi(n)$ except the trusted entity, so no one can forge k'. It means no one can get the legal one-off blind public key $(n_{A,k'}, e_{A,k'}, v_{A,k'}, k')$ by calculating $e_{A,k'} = e_{A,k}^r$ mod n, $v_{A,k'} = v_{A,k}^r$ mod n.

4.5   Unfraudulence

A can't cheat with one-off blind public key. If A cheating, B commits $(n_{A,k}, e_{A,k}, v_{A,k}, k)$ to the trusted entity under the court's grant. Now $\varphi(n) = (p-1)(q-1) = (2p'+1-1)(2q'+1-1) = 4p'q'$, and p', q', $k(k > 2)$ is prime number. The difficulty of k just being p' or q' is equal to the difficulty of factorizing n. Except the case k is p' or q', $gcd(k, \varphi(n)) = 1$ comes into existence. So the trusted entity can calculate $t = k^{-1}$ mod $\varphi(n)$, then can reveal A's identity by calculating $e_{A,k}^t = (e_A^k)^t = e_A$ mod n.

5.   The properties of one-off blind public key protocol

5.1   One transform blind signature

For proofing A has the legal one-off blind public key, it needs the trusted entity's signature on the generative factor of blind public key, then A can generate the blind signature on the part of one-off blind public key based on the generative factor of blind public key, and B can verify it. In our scheme, this blind signature is different with the traditional blind signature [3, 4], only one transform is done, the message and the signature are transforming to another message and the corresponding signature at the same time, we call this blind signature as one transform blind signature. Its process is as follows:

(1) B signs on message m, gets the signature s. B sends (m, s) to A.

(2) A transforms message m to m'. In the same time, A transforms the signature s to s'. Now, A get the signature s' on m'.

In this process, the trusted entity doesn't know the final message m' that will be signed, and the final signature will be verified too, so it is one kind of blind signature. In this kind of blind signature, there are one important different from the traditional blind signature: In the traditional blind signature, A knows the final message that will be signed in advance, but in one transform blind signature, A doesn't know the final message that will be signed in advance. In fact, in one-off blind public key, it signs on the part of the blind public key at the latest, the user's blind public key is generated temporarily when it will be used, and the one-off blind public key is different every time.

5.2   The check on one-off blind public key

(1) B must check A has registered the trusted entity, and the one-off blind public key that A generated bases on the generative factor of blind public key that the trusted entity issues for A. It can be guaranteed through the blind signature on the part of the blind public key that is made by the trusted entity.

(2) B must check A has used the blind public key generative algorithm he promised and the one-off blind public key he published.

If A hasn't used the blind public key generative algorithm he promised and the one-off blind public key he published, and B doesn't check it. It means if A commits with one-off blind public key, no one including the trusted entity can reveal A's identity. In general, the trusted entity won't interfere the communication between A and B, so B must check it, and B can't get any useful information from it.

5.3   The compose of one-off blind public key

From above analysis, we can get the conclusion: one-off blind public key is composed of five parts:

(1) the final public key used for signing or encrypting.

(2) the blind signature on the part of one-off blind public key that is made by the trusted entity.

(3) the blind factor for blind transform in one transform blind signature.

(4) the identifier of the blind public key generative algorithm.

(5) the identifier of the zero-knowledge proof for verifying the validity of one-off blind public key.

In our scheme, the blind public key generative algorithm and the identifier of the zero-knowledge proof for verifying the validity of one-off blind public key are specified by the trusted entity, this point isn't presented directly in one-off blind public key. In fact, the trusted entity can specify more than one blind public key generative algorithm, every blind public key generative algorithm can correspond with one zero-knowledge proof protocol, and the user can select them when one-off blind public key is generated.

5.4   The functions and the rights of the trusted entity

In our scheme, the functions of the trusted entity are as follows:

(1) Generate the generative factor of blind public key for every user.

(2) Reveal the user's identity when he tries to cheat.

It is well known, the cooperation of B and the trusted entity can reveal one user's identity in one communication, but can't view other communications of this user, and it is more difficult to view one fixed user's communication. This point can ensure the user can't cheat with one-off blind public key, and the user's communications have anonymity and disconnection. In our scheme, the trusted entity can forge one-off blind public key and its corresponding private key, but we don't think the trusted entity will do it, otherwise it isn't the trusted entity.

5.5   Comparison with group signature

(1) If treated the trusted entity as the group, only the member of the group can generate one-off blind public key.

(2) The receiver can verify the validity of one-off blind public key, but can't discover the user's identity.

(3) When the disputation occurs, the trusted entity can discover the user's identity with the one-off blind public key.

However, the purpose of we generate one-off blind public key is to use one-off blind public key, and the using of one-off blind public key is to protect user's privacy. The using of one-off blind public key is on behalf of the user, it is independent of the group. For the receivers, they need the trusted entity can ensure the validity of one-off blind public key and can discover the user's identity when the user tries to cheat.

6.   Conclusion

The definition and properties of one-off blind public key are proposed. The analysis of one-off blind public key and its protocol is done. A new scheme is presented, it can generate one-off blind public key based on RSA. In this scheme, when the one-off blind public key is used for encrypting, the session key based on public key is generated.

References

[1] Boneh D, et.al. On the importance of checking cryptographic protocols for faults. *Eurocrypt'97*, 1997.

[2] Chaum D, H. E. V., Group signature. *Eurocrypt'91*, 1991.

[3] D, Chaum, Blind signatures for untraceable payments. *Crypto'82*, 1982.

[4] D, Chaum, Security without identification: Transaction systems to make big brother obsolete. *Commun. of ACM*, 28, 1985.

[5] Gennaro R,et al., Rsa-based undeniable signatures, Journal of Cryptology. 13(4), 2000.

[6] Zhang Giupu, G. B., One-off blind public key based on id. *ACTA ELECTRONICA SINICA*, No.5, 2003.

ANALYSIS ON THE TWO CLASSES OF ROBUST THRESHOLD KEY ESCROW SCHEMES*

Feng Dengguo

State Key Laboratory of Information Security
Institute of Software of Chinese Academy of Sciences
Beijing 100080, China
fdg@is.iscas.ac.cn

Chen Weidong

State Key Laboratory of Information Security
Graduate School of Chinese Academy of Science
Beijing 100039
Institute of Electronics of Chinese Academy of Sciences
Beijing 100080, China
zccwd@yahoo.com.cn

Abstract   In this paper two kinds of "Robust Threshold Key Escrow Schemes" are analysed systematically and many practical attacks on them, such as Subliminal Channel Attack£¬Cheating Attack, etc., are provided. In the meantime, we show whether they gain their end of robustness is worth deliberating. In addition, we also discuss the necessity of some groupware of protocol and establish the basis of analysis on Key Escrow Scheme.

Keywords: key escrow; threshold scheme£»robustness; improved RSA cryptosystem; ElGamal cryptosystem

1.   Introduction

There exist such controversies in the field of information security: How to solve the contradiction between the government's monitoring individual communication(for example it is involved in the security of the country) and protecting privacy. One possible solution is: In communication net, the private keys of

*Supported by the National Gtand for Fundamental Research 973 Program of China under Grant No. G1999035802; the National Natural Science Foundation of China under Grant No. 60253027

users is holden by several escrow agencies; if necessary, for example, gaining the warrant of the court, the monitor can trace one's communication using the shares from escrow agencies. The way has led to a new cryptographic subfields: Key Escrow. Furthermore, NIST(National Institute for Standards and Technology) published EES(Escrowed Encryption Standard) based on hardware chips [1] in 1994.

The current research on Key Escrow mainly focuses on: How to present new secure key escrow schemes not depending on hardware chips, i.e. realizing software key escrow using the public algorithms and protocols.

Since 1990s, the research has made great progress. However, there is still much longer road to perfect key escrow schemes.

In 1995, Desmedt proposed a communication scheme based on public key cryptosystem (ElGamal)[2], which permits the monitor tracing reciver according to LEAF(Law Enforcement Access Field). But Lars R showed soon that Demedt scheme is not secure because it can not prevent several synergic receivers from cheating[3].

In 1995, Shamir put forward the idea of partial key escrow[4] in order to alleviate the loss caused by possible dishonest monitor. In addition, Nechvatal introduced the concept of threshold key escrow[5].

In 1998, Mike Burmster solved the problem of "monitoring once, monitoring forever" which exists in many schemes. The main approach is adding time into secret key of user[6].

We will analyse the results of [7, 8, 9] in this paper. In these papers Cao brings forward two kinds of *Robust Threshold Key Escrow Schemes*(RTKES), where "robustness" refers to the property that even if several malicious escrow agencies work together, it is impossible to recover the secret key escrowed.

We show in our paper that the two schemes mentioned above are not secure and whether they gain their end of robustness is worth deliberating. In fact we are doubtful of the existence of such schemes. In addition, we also discuss the necessity of some groupwares of protocol and establish the basis of analysis on Key Escrow Scheme. For simplicity we denote "Key Escrow System/Scheme" by KES.

2.   Review of two classes of robust threshold Key Escrow Schemes (RTKES)

2.1   RTKES1 based on factorizing

The agencies in RTKES1.

1) KMC(Key Management Center): with responsibility for generating , distributing and managing almost all secret keys.

2) The set of escrow agencies T: holding parts of secret key of public key cryptosystem $E_2(sk, e)$ and helping monitor gain its end.

3)Monitor W: with responsibility for monitoring the communication of users.
The encryption algorithms in RTKES1.

1) $E_1(M, sk)$: A standard block encryption algorithm with session key sk for secure communication of users.

2) $E_2(sk, e)$: So called "improved RSA" used to encrypt session key, which is the main part of LEAF.

3) $S(H(M))$: User's signature, where $H()$ is some hash function.

Description of KES. Assume the private key of $E_2()$ is

$$d = d_1^{-1}d_2 \mod \phi(N)/2,$$

where ϕ is Euler function.

First, KMC divides d_2 into n subkeys using the secret sharing scheme of Shamir. Each $T_i (i = 1, 2, ..., n)$ receives respective share from KMC and can verifies it. Then d_1^{-1} is transmitted to the user U through private channel.

In every communication, sender U and receiver V firstly set up a session key sk using some fixed protocol. Then U transmits the ciphertext with LEAF $c = (E_1(M, sk), LEAF)$ to V, where $LEAF = (E_2(sk, e)^{subd1}, S(H(M)),$ $C(U))$, where $subd1 = d_1^{-1}$, C(U) is the certificate of U. When receiving c, V can recover message M and verifies the signature. If needing monitor, W firstly obtains the information of subkey d_2 from T and can recover the session key sk according to $LEAF$. It should be noted that W can not obtain the private key d. Details can be seen in [7, 8, 9].

2.2   RTKES2 based on DLP(Discrete Logarithm Problems)

The most agencies in RTKES2 is the same as RTKES1, except that:
1) d_1^{-1} is also controlled by monitor W;
2) ElGamal cryptosystem[2] replaces the "Improved RSA".

In addition, RTKES2 also includes a new sub protocol called" monitor authentication protocol" by us, which satisfies the need that W verifies the shares from T.

2.3   Analysis of [7, 8, 9]

[7, 8, 9] argues that there exist robust threshold key escrow schemes, i.e.
1) The schemes RTKES1 and RTKES2 solve the problem "monitoring once, monitoring forever";
2) The two schemes have the property of robustness and are secure against the coalescent of several malicious escrow agencies.
3) Escrow agencies can verify the shares from the KMC.
4) monitor authentication protocol can identify malicious escrow agencies.
5) The two schemes are secure against LEAF feed back attack.

3.   Our viewpoints

We think that RTKES1 and RTKES2 are not secure.

1) KES depend on KMC without measure. In some sense, if W realizes monitor using KMC, the protocols are trivial.

2) "Improved RSA" leaks the lsb of message encrypted.

3) For RTKES1, ite robustness does not consider of the factors of malicious users. And for RTKES2, in fact its robustness is realized by changing W into a special escrow agency with monitor function.

4) The sub protocol that T verify the shares is not necessary under the assumption of [7, 8, 9].

5) The two schemes are not secure against the subliminal channel attack.

6) For RTKES2, if existing malicious users or active (malicious) escrow agencies, monitor authentication protocol is not secure.

4.   Analysis basis on KES

Many analysis methods on KES have been proposed at present. Here, we try to give a normative analysis basis on KES.

Definition 4.1. *(KES) KES is a tuples of five elements as follows: (U,T,W,LMC, V), Where both the number of elements of W and KMC may be 1, V is trusted party. The elements above are linked by cryptographic protocols and satisfy the conditions as follows:*

1) Users carry out communication using LEAF added way;

2) IF all parties obey the rules, W can monitor the communication of users.

The security can be ascribe to the security of sub protocols, such as session key distribution, secure transmission, key escrow, and so on. For the sake of analysis conditions, we introduce the concept of credible degree function.

Definition 4.2. *(credible degree function TR) TR is such a function mapped from all parties of KES to* $[0, 1]$, *where the function value 1 represents believable completely and 0 on the contrary.*

Obviously the values of function TR denote the believable degree of the parties in KES and TR is not computable. But we think that any trusted third party may assess the value of it.

Axiom 1. $TR(U) \leq TR(T) \leq TR(W) \leq TR(KMC)$

Obviously the axiom is reasonable. In fact, $TR(KMC)$ is almost 1 in the two RTKES. So, the security of KES is the security degree under some distribution of TR. The analysis as follows all is based on the axiom.

5. Analysis on RTKES1

5.1 Analysis on Improved RSA

Firstly the Improved RSA is introduced.

$N = pq$ is Blum integer and assume: $gcd(b, N) = 1$, Jacobi symbol $\left(\frac{b}{N}\right) = -1$, $gcd(e, \phi(N)/2) = 1$, $ed = (\phi(N)/4 + 1)/2 \ mod(\phi(N)/2)$. Parameters b, e, N is public and d is secret.

When encrypting message x. If $\left(\frac{x}{N}\right) = 1$, $E(x) = x^{2e}$; otherwise $E(x) = (bx)^{2e}$. The ciphertext is $c = (E(x), c_1, c_2)$, where, if x is even, set $c_1 = 0$, else $c_1 = 1$; if $\left(\frac{x}{N}\right) = 1$, $c_2 = 0$, otherwise $c_2 = 1$. Decrypting is obvious.

Our main result of analysis is as follows and the proof is omitted.

Theorem 5.1. *For Improved RSA, we have:*

1) For any message x_1, x_2, the probability of equation $E(x_1 x_2) = E(x_1) \cdot E(x_2)$ holds true is 0.75 at least.

2) $Parity(y) = c_1$, i.e. the least significant bit of message is leaked.

5.2 Analysis on escrow protocol

This is a sub protocol on distribution of shares generated by KMC. As protocol finished, each escrow agency obtains the legal shares of secret key of user and each share can be verified.

We note that this is a non-interactive protocol and the information is transmitted through private secure channel([7, 8, 9]). So it is not easy to see that the only possible attack on the protocol is the cheating attack of KMC (It should be noted that the security goal of the protocol is escrow agencies can obtain their shares correctly.) However, by axiom 1, we have known that it is impossible to imagine that KMC is dishonest. Therefore, we can give the conclusion: authentication groupware of the protocol is not necessary. In trivial sense, the security of the protocol can be proved.

5.3 Subliminal channel attack on communication protocol

According to introduction of RTKES1, it can be seen that:

1) Receiver V does not need $LEAF$ indeed when decrypting because the session key sk is only set up by U and V.

2) The part of private key of U, $subd1 = d_1^{-1}$, is not escrowed, i.e. $subd1$ is still controlled by U.

So, U and V can attack on it in the way as follows.

U may sends his message to receiver V in the way:

$c' = (E_1(M, sk), LEAF)$, $LEAF = (E_2^{subd1}(sk', e), S(H(M')), C(U))$

Where $sk' = h(sk)(h(\ )$ is some public secure one-way function, such as hash function), $M' = E_1(E_1(M, sk), sk')$ and assume the encrypting algorithm and decrypting one are same.

U can also send his message in such way: $c' = (E_1(M, sk), LEAF)$, $LEAF = (E_2^{subd1'}(sk, e), S(H(M'')), C(U))$ where $subd1' = d_1^{-1}h(d_1)$, $h(d_1)$ may be public and $M'' = E_1(E_1(M, sk), sk^{h(d_1)})$.

We can see easily that in case of monitoring, W can only obtain false session key $sk' = h(sk)$ or $sk^{h(d_1)}$ from the escrow agencies and even if W verifies the signature in LEAF, it can not detect the existence of subliminal channel. In fact, the true session key of U and V is sk, but under the assumption of one-way function or factorization problem, W can not derive sk from sk'.

Theorem 5.2. *RTKES1 can not resist the subliminal channel attack launched by malicious users.*

Considering of least value of $TR(U)$, the conditions needed by such attack is very weak.

5.4   Analysis of monitor protocol

Theorem 5.3. *In case of cahoot of malicious users and k escrow agencies, integer N can be factorized.*

The proof is easy and be omitted. In fact, considering of least value of $TR(U)$, it is not suitable to take U as one escrow agency in order to solve the problem of depending on escrow agencies without measure.

6.   Analysis on RTKES2

As we know, RTKES2 is similar to RTKES2 except that encrypting algorithm adopts ElGamal. Therefore the analysis conclusions above also hold true. However, there exist another subliminal channel attack for both RTKES1 and RTKES2 and monitor protocol of the latter is not secure too.

6.1   Subliminal channel attack based on signature schemes

Here, we want to show the point that it is dangerous to prevent subliminal channel attack only by signature schemes.

According to [14, 15], there exists subliminal channel in ElGamal signature schemes.

Here we construct one kind of subliminal channel using the signature scheme based on improved RSA: Sender U leaks his private key for signing to receiver V on purpose previously, where the private key for signing $d = sk + m \, mod\phi(N)$. And They use $h(sk)$ as their session key where $h(\ )$ is a secure hash function. Then U sends his message in the way as follows:

$c = (E_1(M, h(sk)), LEAF)$, $LEAF = (E_2^{subd1}(h(sk), e), S(H(M))$

As a result, in case of monitor, W can only obtain the $h(sk)$ form the escrow agencies, i.e. recover some disguised message M. Because of one-wayness of

$h(\ )$, W can not derive sk from $h(sk)$. On the other hand, V can easily recover the intended message m by computing $m = (d - sk) \mod \phi(N)$. It should be noticed that sk is set up only by U and V in both RTKES1 and RTKES2.

6.2 Cheating attack on monitor protocol

Firstly we give a concise description of monitor protocol of RTKES2. The sub protocol is devised to make W verify the shares from escrow agencies, i.e. W should have the ability of identifying malicious escrow agencies.

Parameters. KMC publicizes the public key of U $y = g^d$, where private key $d = d_1^{-1}d_2$, d_1^{-1} is sent to W through private channel previously , d_2 is divided into n shares $b_1, ..., b_n$ by secret sharing scheme and these shares are given to escrow agencies. In case of monitoring, escrow agency $T_i(i = 1, 2, ..., n)$only refers $Q_i = u^{b_i} = g^{kb_i}$to W, where g^k is the first part of ciphertext of sk (Note that we adopts ElGamal public key algorithm to encrypt session key).

Protocol. For $i = 1, 2, ..., n$, the steps as follows. 1) T_i sends $Q_i = u^{b_i}$ to W; 2) W chooses α_i and β_i in F_q randomly and computes $w_i = Q_i^{\alpha_i}y_i^{\beta_i}$. Then sends it to T_i; 3) T_i computes $R_i = w_i^{b_i^{-1}}$ and sends it to W; 4) W verifies if the equation $R_i = u^{\alpha_i}g^{\beta_i}$ holds true. If yes, W accepts the shares, otherwise, W alleges T_i is dishonest.

Analysis. It should be noted that the verification equation is involved in $u = g^k$. We think it is not suitable because $u = g^k$ is generated by user U whose value of $TR(U)$ is smallest. So two cheating protocols can be constructed as follows.

Case1. Some malicious sender U replaces $u = g^k$ used in $LEAF$ with any g^l and l is given to T_i. In case of monitoring, some malicious T_i can do so in the step 1) of the protocol: T_i computes $a_i = b_i l$ and sends $Q_i = g^{a_i}$ (instead of g^{kb_i}) to W. On the other hand, in step 3) T_i answers the challenge of W with $R_i = w_i^{b_i^{-1}}$. The readers can verify easily that by doing so, T_i can cheat W with $Q_i = g^{lb_i}$. Note that the true share is $Q_i = g^{kb_i}$.

Case2. The malicious escrow agency T_i is active and replaces $u = g^k$ in $LEAF$ with g^l chosen by itself. Note that this whould not affect the normal communication of users.

The reason for the success of cheating attack is that the communication channel of users is not secure, i.e. there are not private channel like KMC. In addition, the attack conditions are feasible according to axiom 1.

6.3 Analysis on Robustness

We can see that in fact RTKES2 realizes the robustness by regarding W as an agency with functions of escrow and monitor. Obviously this strengthens the rights of W greatly. Considering the case of dishonest W[4], the method may be not worth the candle.

7.   Tag

Considering separately, we think that in some sense [7, 8, 9] solve the problem "Monitoring once, Monitoring forever". However, There exist some deficiencies: Session key sk is chosen completely by users and master keys is escrowed in common ways. Therefore the method does not solve the problem above in the complete sense. We argue that the problem can be solved using key escrow scheme with Limited time span [6].

References

[1] National Institute for Standards and Technology , Escrowed Encryption Standard, Federal Information Processing Standards Publication 185, U.S. Dept. of Commerce , 1994.

[2] TAHER EIGAMAL, A Public Key Cryptosystem and a Signature Scheme Based on Discrete Logarithms, IEEE TRANS. On Information Theory, vol IT-31, NO.4,1985.

[3] Lars R.Knudsen, Torben P.Pedersen, On the difficulty on software key escrow, EURO-CRYPT'96, p237-244.

[4] S.Micali and R.Shamir, Partial Key Escrow, February 1996.

[5] Nechvatal J.A, Public-key-based key escrow system, Journal of Systems Software, 1996, 35(1): 73-83.

[6] Mike Burmster, Y.Desmedt and Jennifer Seberry, Equitable Key Escrow with Limited time span, AsiaCrypto'98, 1998.

[7] Cao Zhen-Fu, Two classes of Robust Threshold Key Escrow Schemes. Journal of Software, Vol.14, No.6, 2003.

[8] Cao Zhen-Fu, Li Jiguo, A Threshold Key Escrow Scheme Based on ElGamal Public Key Cryptosystem, Chinese Journal of Computers. 2002, 25(4):346-350.

[9] Cao Zhen-Fu, A Threshold Key Escrow Scheme based on public key cryptosystem. Science in China(Series E), 2001, 44(4):441-448.

[10] Rivest,R.L., Shamir,A. and Adleman, L., A method for Obtaining Digital Signatures and Public-key Cryptosystem, Comm.ACM Vol.21(2), pp.120-126,FEB.1978.

[11] Shamir A., How to share a secret, Communications of the ACM, 1979, 22(11):612-613.

[12] Fen Dengguo, Pei Dingyi, Introduction of Cryptology, Science Press, 1999.

[13] J.Kiliam, T.Leighton, Fair Cryptosystems, Revisited , AsiaCrypto'96, 1996.

[14] Simmons,G.J., The Subliminal Channel and Digital Signatures, Advance in Cryptology-Eurocrypto'84, Springer-Verlag, 1985,pp.51-67.

[15] Feng Dengguo etal.£¬Review of foreign key escrow systems£¬Technical Report£¬DCS Center, Chinese Academy of Sciences£¬Beijing£¬Nov., 1997.

PRIVACY-PRESERVING APPROXIMATELY EQUATION SOLVING OVER REALS[*]

A multi–party protocol to solve equations

Zhi Gan, Qiang Li, Kefei Chen
Department of computer science and engineering
Shanghai JiaoTong University
Shanghai 200030, China
{gan-zhi, chen-kf}@cs.sjtu.edu.cn
qiangl@sjtu.edu.cn

Abstract   Equation solving is a important scientific computations that's generally employed. Solutions to this problem are widely used in many areas such as banking, manufacturing, electric engineering and telecommunications. However, the existing solutions do not extend to the privacy–preserving cooperative computation situation, in which the equations are shared by multiple parties, who do not want to disclose their data to other parties.

In this paper, we formally define these specific privacy–preserving cooperative computation problems, and present protocols to solve them. Besides, a new multi–party protocol to handle computation over reals is presented.

Keywords: secure multi–party computation, equation solving

1.   Introduction

The rapid development of distributed systems raised the natural question of what tasks can be securely performed. With the application of multi–party computation, cooperative computation could occur between mutually untrusted parties, or even between competitors.

In this paper, we introduce the privacy–preserving cooperative equation solving(PPCES) problem. The general definition of the PPCES problem is that two or more parties want to solve an equation based on their private inputs, but neither party is willing to disclose its own input to anybody else(including a so–called trusted third party).

[*] This work was supported by NSFC under grants 90104005, 60273049.

Generally speaking, a secure multi–party computation problem deals with computing a function on any input, in a distributed network where each participant holds one of the inputs, ensuring that no more information is revealed to a participant in the computation than can be computed from that participant's input and output. The history of the multi–party computation problem is extensive since it was introduced by Yao [2]and extended by Goldreich, Micali, and Wigderson[8], and by many others.

In theory, the general secure multi–party computation problem is solvable using circuit evaluation protocol[2, 8, 7]. In the circuit evaluation protocol, each functionality F is represented as a Boolean circuit, and then the parties run a protocol for every gate in the circuit. While this approach is appealing in its generality, the communication complexity of the protocol it generates depends on the size of the circuit that expresses the functionality F to be computed, and in addition, involves large constant factors in their complexity. Therefore, as Goldreich points out in [7], using the solutions derived by these general results for special cases of multi–party computation can be impractical; special solutions should be developed for special cases for efficiency reasons[10, 4]. This is our motivation of seeking special solutions to equation solving problems, solutions that are more efficient than the general theoretical solutions.

There are several ways to share an equation. Depending on how such an equation is shared by Alice and Bob, or in another word how Alice and Bob cooperate with each other, the problems could appear in a variety of forms. Table 1 describes two different types of cooperation.

Table 1.   Various ways of cooperation

	Case 1	Case 2
Alice's share	$f(x)$	$f(x)$
Bob's share	$g(x)$	$g(x)$
target equation	$h(f(x), g(x)) = 0$	$f(g(x)) = 0$

2.   Approximately Multi–party Computation over Reals

All we known existed methods deal with computation in a finite field[3, 2, 1, 7, 9]. We now present a multi–party protocol over reals based on *verifiable secret sharing scheme*[5]. Every function over reals of n inputs can be efficiently computed by a complete network of n participates. And if no faults occur, no set of size $t < n/2$ of players gets any additional information. Even if Byzantine faults are allowed, no set of size $t < n/3$ can either disrupt the computation or get additional information. As Gennaro points out in [6], multi–party computation can be efficiently performed in $GF(p)$. Then a natural idea is to find a way to code real number into a element of $GF(p)$ and decode field element to real number, then we can perform approximate computing over reals. For example, we can employ following conversion rule between field element F and real number R:

1 Select an adequate prime p which is large enough to serve following computations. Public parameter p should satisfy the condition that $p/2$ is greater than any of the inputs, temporary results and final results.

2 According to required accuracy, Alice and Bob select another public parameter k as the conversion rate between real number and field element. For example, if error should be controlled in the range of $\pm 0.5 \times 10^{-5}$, k should be 10^6.

3 Using following function to encode real number:

$$F = E(R) = \begin{cases} [R \times k], (R \geq 0) \\ p + [R \times k], (R < 0) \end{cases} \quad (1)$$

where $[\ ]$ means the nearest integer function.

4 Using following function to decode:

$$R = D(F) = \begin{cases} F/k, (F \leq p/2) \\ -(p - F)/k, (F > p/2) \end{cases} \quad (2)$$

After conversion, three operations–addition, subtraction and multiplication– can be securely conducted in finite field[5]. When multiplication is mixed with other operations, then conversion rate should be adjusted. For example, if we want to securely compute $a + b \times c$, we compute $a \times k^2 + (b \times k)(c \times k)$ over finite field.

What we should think about carefully is division operation. Division over reals is completely different than over finite field. Assume a and b are shared by Alice and Bob, then a/b can be computed as following:

1 Alice selects random number x_1 and shares it with Bob. Bob selects random number x_2 and shares it with Alice.

2 Alice and Bob now jointly compute and publish $m = b \cdot x_1 \cdot x_2$.

3 Since m is published, Alice and Bob can compute and share $\frac{1}{m}$.

4 Alice and Bob now jointly compute $r = a \cdot \frac{1}{m} \cdot x_1 \cdot x_2$, which is the final result.

There is a problem in above protocol. Since $m = b \cdot x_1 \cdot x_2$ is published, then b must be a factor of m. Therefore Alice and Bob get some information about *b.* It's a violation to security requirements of secure multi–party computation. We provide a modified version to deal with this problem.

1 Alice selects two random number x_1, ϵ_1 and shares it with Bob. And Bob selects two random number x_2, ϵ_2 and shares it with Alice. ϵ_1 and ϵ_2 are real numbers in the range of $\pm k^{-1}/10$.

2 Alice and Bob now jointly compute and publish $m = b \cdot x_1 \cdot x_2 + \epsilon_1 + \epsilon_2$.

3 Since m is published, Alice and Bob can compute and share $\frac{1}{m}$.

4 Alice and Bob now jointly compute $r = a \cdot \frac{1}{m} \cdot x_1 \cdot x_2$, which is the final result.

In above protocol, ϵ_1 and ϵ_2 are small enough to keep the accuracy of calculation. The new protocol is secure and can be easily modified to serve any number of participates. Here is some sample text.

3.   Secure Multi–Party Equation Solving Problems and Protocols

Newton–Raphson Method is a root–finding algorithm which uses the first few terms of the Taylor series of a function $f(x)$ in the vicinity of a suspected root to zero in on the root. For $f(x)$ a polynomial, Newton's method is essentially the same as Horner's method. The Taylor series of $f(x)$ about the point $x = x_0 + \epsilon$ is given by

$$f(x_0 + \epsilon) = f(x_0) + f'(x_0)\epsilon + \frac{1}{2}f''(x_0)\epsilon^2 + \dots \qquad (3)$$

Keeping terms only to first order,

$$f(x_0 + \epsilon) \approx f(x_0) + f'(x_0)\epsilon \qquad (4)$$

This expression can be used to estimate the amount of offset ϵ needed to land closer to the root starting from an initial guess x_0. Setting $f(x_0 + \epsilon) = 0$ and solving (4) for $\epsilon \equiv \epsilon_0$ gives

$$\epsilon_0 = -\frac{f(x_0)}{f'(x_0)} \qquad (5)$$

which is the first–order adjustment to the root's position. By letting $x_1 = x_0 + \epsilon_0$, calculating a new ϵ_1, and so on, the process can be repeated until it converges to a root using

$$\epsilon_n = -\frac{f(x_n)}{f'(x_n))}. \qquad (6)$$

When the method converges, it does so quadratically.

In a multi-party environment, after selecting adequate x_0, we can solve equation using Newton-Raphson method if we can compute out $f(x)$ and $f'(x)$ on arbitrary x. In both the two cases listed in Table 1, we can easily compute the function results and their derivatives.

In the first case, Alice has $f(x)$, Bob has $g(x)$, and they want to solve equation $h(f(x), g(x)) = 0$, where h is a function of two inputs and uses only addition, subtraction, division, and multiplication operations. And $h(f(x), g(x))$ can be securely computed as following:

$$\frac{d\, h(f, g)}{d\, x} = \frac{\partial h}{\partial f} \cdot \frac{d\, f}{d\, x} + \frac{\partial h}{\partial g} \cdot \frac{d\, g}{d\, x} \tag{7}$$

where Alice can privately computes $\frac{\partial h}{\partial f} \cdot \frac{d\, f}{d\, x}$ and Bob can privately computes $\frac{\partial h}{\partial g} \cdot \frac{d\, g}{d\, x}$. Then we can use Newton–Raphson method to solve the equation $h(f(x), g(x)) = 0$.

In the second case, Alice has $f(x)$, Bob has $g(x)$, and they want to solve equation $f(g(x)) = 0$ where $f(x)$ is a polynomial function. Let $r(x) = f(g(x))$. Since we limit $f(x)$ to be a polynomial function, $r(x)$ and $r'(x)$ can be securely computed out. And $r'(x)$ can be computed as following:

$$r'(x) = f'(g(x)) \cdot g'(x) \tag{8}$$

After initial value of x is selected, participants can perform multi–party computation continuously until the approximate result is found. The whole computation process can be divided several iterations. And each iteration is composed of two phases: a checking phase and a computing phase. In checking phase, all participates check together if $f(x_i)$ resides in $(-1/k, 1/k)$. If the answer is true, then x_i is the approximate root of equation. Else, the computing phase is performed. In this phase, all participates compute together to get x_{i+1}.

For some equations, Newton–Raphson method can't converge into right answer. We can employ other methods such as bisection method to deal with such situation. In fact, we found bisection method can deal with much more kinds of one–dimensional equations than Newton–Raphson method in multi–party computation scenario. But bisection method has shortcomings too. It's well–known that bisection method can't be used to resolve multi–dimensional problem, and it's much slower than Newton–Raphson method. So we still choose Newton–Raphson method as the default one.

4. Summary and Future Work

In this paper, we provide a new multi–party computation protocol over reals. Through encoding real number into member of a finite field, addition, subtraction and multiplication can be done in a very simple way. And we can easily know that the addition, subtraction, and multiplication protocols are as secure as Gennaro's protocol[6]. The division is done in a much complexer way.

We also studied the problem of equation solving in a cooperative environment where neither of the cooperating parties wants to disclose its private data to the other party. Our preliminary work has shown that this problem, the secure multi–party equation solving problem, could be solved in a way more efficient than the general circuit evaluation approach.

Apart from those basic equations studied in this paper, many other types of equations are also used in practice. A future direction would be to study more complicated equation solving problems.

References

[1] Abe, M. (1999). Mix-networks on permutation networks. In *ASIACRYPT'99*, pages 258–273.

[2] A.C.Yao (1982). Protocols for secure computations. In *In Proc. 23rd IEEE Symp. On the Foundation of Computer Science*, pages 160–164. IEEE.

[3] Cramer, R., Damgård, I., and Maurer, U. (2000). General secure multi-party computation from any linear secret-sharing scheme. *Lecture Notes in Computer Science*, 1807:316–??

[4] Du, W. and Atallah, M. J. (2001). Privacy-preserving statistical analysis. In *Proceedings of the 17th Annual Computer Security Applications Conference*, pages 102–110, New Orleans, Louisiana, USA.

[5] Feldman, P. (1987). A practical scheme for non-interactive verifiable secret sharing. In *28th FOCS*, pages 427–437.

[6] Gennaro, R., Rabin, M., and Rabin, T. (1998). Simplified VSS and fast-track multiparty computations with applications to threshold Cryptography. In *Proceedings of the 1998 ACM Symposium on Principles of Distributed Computing*, pages 101–111.

[7] Goldreich, O. (2000). Secure multi-party computation. Working Draft.

[8] Goldreich, O., Micali, S., and Wigderson, A. (1987). How to play any mental game. In *In Proceedings of the 19th Annual ACM Symposium on Theory of Computing*, pages 218–229.

[9] Matthew, Franklin, and Habert, S. (1996). Joint encryption and message–efficient secure computation. *Journal of Cryptology*, 9(4):217–232.

[10] W.Du and Atallah, M. J. (2001). Privacy-preserving cooperative scientific computations. In *14th IEEE Computer Security Foundations Workshop*, pages 273–282, Nova Scotia, Canada.

AN AUTHENTICATED KEY AGREEMENT PROTOCOL RESISTANT TO DOS ATTACK

Lu Haining
Dept. of Computer Science and Engineering
Shanghai Jiao Tong University
longsky@263.net

Gu Dawu
Dept. of Computer Science and Engineering
Shanghai Jiao Tong University
gu-dw@cs.sjtu.edu.cn

Abstract   The Authenticated Key Agreement with Key Confirmation protocol proposed by Blake-Wilson et al improves the original Diffie-Hellman key agreement protocol and defeats the man-in-the-middle attack. But it is vulnerable to a Denial-of-Service (DoS) attack, because the responder must perform heavy modular exponential operations before he becomes sure about the identity of the initiator. A modification which forces the initiator to perform modular exponentiation first is presented in this paper. According to the analysis, it can defeat the DoS attack successfully, and provide mutual key authentication and key confirmation as well.

Keywords: authenticated key agreement protocol, denial-of-service attack.

1.   Introduction

The first and best known key agreement protocol is Diffie-Hellman Key Exchange[1], from which many of the commonly used key agreement protocols are put forward. But the original Diffie-Hellman protocol remains following two critical problems:

1   Both sides of the communicating parties can not get an assurance that no other parties can possibly compute the keying information agreed.

2   Each side of the communicating parties can not assure that the party he is communicating with has actually computed the agreed key.

So, this protocol is vulnerable to man-in-the-middle attack. In 1997, Blake-Wilson et al modified the original Diffie-Hellman protocol, and proposed an authenticated key agreement protocol with key confirmation[2](in short, AKAKC protocol), which solved the problems mentioned above. But it is still vulnerable to DoS attack because the responder should perform heavy modular exponential operations before identifying the initiator.

The remainder of the paper is organized as follows. First describes the AKAKC protocol, and presents the basic principle of DoS attack. Then proposes an improved protocol which can defeat DoS attack and provides mutual key authentication and confirmation as well. Finally gives the feasibility and security analysis of the improved protocol.

2. AKAKC Protocol

Figure 1. shows the details of the AKAKC protocol. The notations are described as follows: A and B are trusted two parties who communicate with each other. p is a large prime, q is a large prime divisor of $p-1$, g is an element of order q in Z_p^*. a and b are static private keys of A and B, Y_A and Y_B are static public keys of A and B. H_1 and H_2 are two independent hash functions. MAC is a message authentication code algorithm.

Figure 1. AKAKC Protocol

1 A selects $x \in_R [1, q-1]$ and sends $R_A = g^x$ and $Cert_A$ to B.

2 After receiving requesting message,

    (a) B verifies that $1 < R_A < p$ and $(R_A)^q \equiv 1 \pmod p$. If any check fails, then B terminates the protocol run with failure.

    (b) B selects $y \in_R [1, q-1]$, and computes $R_B = g^y$, $k' = H_1((Y_A)^b \parallel (R_A)^y)$, $k = H_2((Y_A)^b \parallel (R_A)^y)$, and $m_B = MAC_{k'}(2, B, A, R_B, R_A)$.

    (c) B sends R_B, $Cert_B$, and m_B to A.

3 After receiving responding message,

    (a) A verifies that $1 < R_B < p$ and $(R_B)^q \equiv 1 \pmod{p}$. If any check fails, then A terminates the protocol run with failure.

    (b) A computes $k' = H_1((Y_B)^a \| (R_B)^x)$ and $m'_B = MAC_{k'}(2, B, A, R_B, R_A)$, and verifies $m'_B = m_b$.

    (c) A computes $m_A = MAC_{k'}(3, A, B, R_A, R_B)$, $k = H_2((Y_B)^a \| (R_B)^x)$, and sends R_A and m_A to B.

4 B computes $m'_A = MAC_{k'}(3, A, B, R_A, R_B)$ and verifies that $m'_A = m_A$.

5 The session key is k.

This protocol provides mutual key authentication and key confirmation.

3.   DoS attack

To generate a valid requesting message, the initiator only need to choose a number between 1 and p which to the power of q is 1 modulo p. But when receiving a valid requesting message, the responder must perform three modular exponential operations: g^x, g^{ab}, g^{xy}, which cost a lot. If an attacker sends a large amount of valid requesting messages to some party, the target will get its computational resources exhausted soon and can not respond the request from other users. Now the attacker wins.

4.   An improved protocol which can defeat DoS attack

4.1   Basic idea of the improved protocol[3]

The reason why DoS attack works is that it is much more easier to generate the requesting message than the responding one. So we have the following idea. After receiving a valid requesting message, the responder do not perform any heavy computation such as modular exponentiations. Instead, he generates a random fresh material from some secret information and his private key, and pass it to the initiator. Reconstruction of the secret information from this material requires heavy computation. The responder will continue the protocol and perform modular exponentiation only after assuring that the initiator has already reconstructed the secret. Thus, the attacker will fall in heavy computation with the responder together and fail the attack.

4.2   Description of the improved protocol

Figure 2. shows the details of the improved protocol. The notations are described as follows: A and B are trusted two parties who communicate with each other. p is a large prime, q is a large prime divisor of $p - 1$, g is an element of order q in Z_p^*. a and b are static private keys of A and B, Y_A and Y_B are static public keys of A and B. H_1 and H_2 are two independent hash functions. MAC is a message authentication code algorithm.

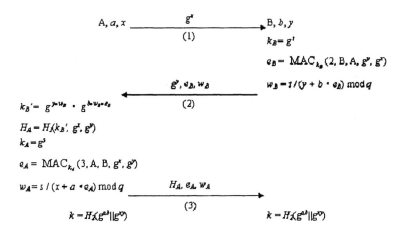

Figure 2.   The Improved Protocol

1 During the precomputation stage, B selects t, y pairs from $[1, q - 1]$ continuously and computes g^t and g^y.

2 A selects $x \in_R [1, q - 1]$, and sends $R_A = g^x$ and $Cert_A$ to B.

3 After receiving message(1):

   (a) B verifies that $1 < R_A < p$ and $(R_A)^q \equiv 1 \ (mod \ p)$. If any check fails, then B terminates the protocol run with failure.

   (b) B selects a (g^t, g^y) pair generated during the precomputation stage, lets $k_B = g^t$, and computes $e_B = MAC_{k_B}(2, B, A, g^y, g^x)$ and $w_B = t \ / \ (y + b \cdot e_B) \ mod \ q$.

   (c) B sends $R_B = g^y$, e_B, w_B and $Cert_B$ to A.

4 After receiving message(2),

   (a) A verifies that $1 < R_B < p$ and $(R_B)^q \equiv 1 \ (mod \ p)$. If any check fails, then A terminates the protocol run with failure.

   (b) A computes $k'_B = g^{y \cdot w_B} \cdot g^{b \cdot w_B \cdot e_B}$ and $e'_B = MAC_{k'_B}(2, B, A, g^y, g^x)$, and verifies that $e'_B = e_B$.

(c) A selects $s \in_R [1, q-1]$, and computes $h_A = H_1(k'_B, g^x, g^y)$, $k_A = g^s$, $e_A = MAC_{k_A}(3, A, B, g^x, g^y)$ and $w_A = s / (x + a \cdot e_A) \bmod q$.

(d) A sends h_A, e_A and w_A to B.

5 After receiving message(3),

(a) B computes $h'_A = H_1(k_B, g^x, g^y)$, and verifies that $h'_A = h_A$.

(b) B computes $k'_A = g^{x \cdot w_A} \cdot g^{a \cdot w_A \cdot e_A}$ and $e'_A = MAC_{k'_A}(3, A, B, g^x, g^y)$, and verifies $e'_A = e_A$.

6 The session key is $k = H_2(g^{ab} \| g^{xy})$.

4.3 The analysis of the improved protocol

1 Against DoS attack

First we suppose that we can limit the requesting messages from the attacker by Network Ingress Filter[4] to the extent that the (g^t, g^y) pairs generated during the precomputation stage won't be exhausted. We also suppose that the amount of the attacker's computational resources are similar to those of the responder's.

After receiving a request, the responder only need to perform a MAC computation and some simple modular arithmetics. After receiving message(3), the responder first computes a hash value h'_A, and verifies that $h'_A = h_A$. If the verification succeeds, it means that the initiator has already computed k_B by performing modular exponentiation. Only after successful verification, will the responder continue the protocol and perform modular exponentiation to further verify the message and compute session key. So, if a malicious attacker wants the responder to perform last modular exponentiation, he must perform heavy computation first when receiving message(2). And he will fall in heavy computation with the responder together, and fail the attack.

2 Provide mutual key authentication

If someone wants to compute the session key $k = H_2(g^{ab} \| g^{xy})$, he must also acquire a or b and x or y besides g^x, g^y, g^a and g^b which can be acquired from the communication messages. Any party except A and B can only compute from e_A, e_B, w_A, w_B and h_A. e_A and e_B are authentication code, h_A is hash value, so they won't reveal the original information. In the expression of w_A and w_B, there are three unknowns (t, y, b) and (s, x, a) respectively. If q is very large, there will be a lot of suitable solutions. So, when the protocol finishes, A and B both can assure that no other party can compute the session key k.

3 Provide mutual key confirmation

In the fourth step, A can determine if B knows g^x by verifying $e'_B = e_B$. Similarly, B can determine if A possesses g^y by verifying $e'_A = e_A$ in the fifth step. And because A and B can get each other's public key, both of them can assure that the other side can compute the session key $k = H_2(g^{ab} \parallel g^{xy})_i$£

5.   Summary

The AKAKC protocol proposed by Blake-Wilson et al is vulnerable to DoS attack because the responder must perform heavy modular exponential computation before identifying the initiator. An improved protocol presented in this paper can defeat the DoS attack by forcing the initiator to perform modular exponentiation first. When attacker launches DoS attack, he will fall in heavy computation with the responder together and fail the attack.

According to the analysis, the improved protocol can defeat the DoS attack successfully, and provide mutual key authentication and confirmation as well.

The idea of designing this protocol can also be applied in revising other public-key based authentication key agreement protocol, such as IKE.

References

[1]   W. Diffie and M.E. Hellman, "New directions in cryptography", *IEEE Trans. Info. Theory* IT-22 No.6 (Nov. 1976), 644-654.

[2]   S. Blake-Wilson, D. Johnson and A. Menezes, "Key agreement protocols and their security analysis", *Proceedings of the sixth IMA International Conference on Cryptography and Coding,* LNCS 1344, 1997, 30-46.

[3]   K. Matsuura and H. Imai, "Protection of authenticated key-agreement protocol against a denial-of-service attack", *Proc. 1998 International Symposium on Information Theory and Its Applications (ISITA '98),* pp.466-470, Mexico City, Mexico, Oct. 1998

[4]   P. Ferguson and D. Senie, "Network Ingress Filtering: Defeating Denial of Service Attacks which employ IP Source Address Spoofing", rfc2267, January 1998.

A COMMENT ON
A MULTI-SIGNATURE SCHEME*

ZHENG Dong, CHEN Kefei
Department of Computer Science and Engineering
Shanghai Jiaotong University, Shanghai 200030, China
(zheng-dong,chen-kf)@cs.sjtu.edu.cn

HE Liangsheng
Stat Key Labs. of Information Security
Graduate School of Chinese Academy of Science
Beijing 100039, China

Abstract   We will show that the structured ElGamal-type multisignature scheme due to Burmester *et al.* be not secure if the adversary attacks key generation.

Keywords: cryptanalysis, multi-signature, authentication

1.   Introduction

Multisignature scheme realizes that plural users generate the signature on a message, and that the signature is verified. Recently, Burmester et al.[1] presented a structured ElGamal-type scheme (Burmester et al.'s scheme), which is based on discrete logarithm problem (DLP). This letter shows that the Burmester et al.'s scheme is not secure if the adversary attacks key generation. In the following, the brief review of Burmester et al.'s scheme is given, and then an attack is proposed.

2.   Brief review of Burmester *et al.*'s scheme

We assume that n signers $I_1, I_2, \ldots, I_n$ generate a signature on a fixed message M according to order fixed beforehand. Burmester *et al.*'s scheme is as following:

* Partial funding provided by NSFC under grants 90104005, 60173032 , 60273049 and 60303026 .

Key generation: In Burmester *et al.*'s scheme, there are three public system parameters. The parameters p and q are two large prime numbers, $p>q$, the parameter $g \in Z_p^*$, is an element with order q. $h()$ is a public hash function. Each user selects his private key $a_i \in Z_p^*$, then computes his public key sequentially as follows:

$$y_1 = g^{a_1} \mod p, \qquad y_i = (y_{i-1}g)^{a_i} \mod p$$

then a public key of ordered group $I_1, I_2, \ldots, I_n$ is set to $y = y_n$.

Signature generation:

1 Generation of r:

   signers $I_1, I_2, \ldots, I_n$ generate r together as follows:

   (a) Player I_1 selects $k_1 \in Z_p^*$ randomly and computes $r_1 = g^{k_1}$ mod p, if $gcd(r_1, q) \neq 1$, then select new k_1 again.

   (b) For $i \in 2, 3, \ldots n$, I_{i-1} sends r_{i-1} to I_i, and I_i selects $k_i \in Z_q^*$ randomly and computes $r_i = r_{i-1}^{a_i} g^{k_i} \mod p$, if $gcd(r_i, q) \neq 1$, then select new k_1 again.

   (c) $r = r_n$

2 Generation of s: Signers $I_1, I_2, \ldots, I_n$ generate s together as follows:

   (a) I_1 computes $s_1 = a_1 + k_1 r H(r, M) \mod q$

   (b) For $i \in 2, 3, \ldots n$, I_{i-1} sends s_{i-1} to I_i. I_i verifies that

   $$g^{s_{i-1}} = y_{i-1} r_{i-1}^{rh(r,M)} \mod p,$$

   then computes

   $$s_i = (s_{i-1} + 1)a_i + k_i r H(r, M) \mod q$$

   (c) $s = s_n$

3 The multisignature on M by order $(I_1, I_2, \ldots, I_n)$ is given by (r, s).

Signature verification:
A multisignature (r, s) on M is verified by

$$g^s = yr^{rh(r,M)} \mod p \tag{1}$$

If the adversary attacks key generation, the above scheme is not secure at all.

3. Our attack

Key generation is the same as Burmester et al.'s scheme but that player I_j is bad and generates his public key by choosing a secret key $a_j \in Z_q^*$ and setting $y_j = g^{a_j} \mod p$. The key of ordered group $(I_1, I_2, \ldots, I_n)$ is set to $y = y_n$.

In this case, The multisignature (r, s) on M can be generate without $I_1, \ldots, I_{j-1}$ signing it:

1 Generation of r:

(a) Player I_j selects $k_j \in Z_q^*$ randomly and computes $r_j = g^{k_j} \mod p$, if $gcd(r_j, q) \neq 1$,then select new k_j again.

(b) for $i \in j+1, \ldots, n$, a signer I_{i-1} sends r_{i-1} to I_i, and I_i selects $k_i \in Z_q^*$ randomly and computes $r_i = r_{i-1}^{a_i} g^{k_i} \mod p$, if $gcd(r_i, q) \neq 1$,then select new k_i again.

(c) $r = r_n$

2 Generation of s: signer $I_1, I_2, \ldots, I_n$ generate s as follows:

(a) I_j computes $s_j = a_j + k_j rh(r, M) \mod q$

(b) for $i \in j+1, \ldots, n$, I_i sends s_{i-1} to I_i, I_i verify that

$$g^{s_{i-1}} = y_{i-1} r_{i-1}^{rh(r,M)} \mod p,$$

then computes

$$s_i = (s_{i-1} + 1)a_i + k_i rh(r, M) \mod q$$

(c) $s = s_n$

3 The bad multisignature on M is (r, s).

Verification: it is obvious that for the above bad multisignature (r, s), equation(1) is still hold:

$$g^s = yr^{rh(r,M)} \mod p$$

The above attack shows that I_j can cheat $I_{j+1}, \ldots, I_n$ to sign any message M without knowing $I_1, \ldots, I_{j-1}$ not signing it. Especially, when $j = n$, player I_j can sign any message M it wants on behalf of the entire group $I_1, I_2, \ldots, I_n$.

4. Summary

we have presented an attack on Burmester et al.'s scheme, the attack shows that Burmester et al.' scheme is insecure against attacks on key generation. It is possible to modify the Burmester et al.'s scheme by requiring that each player I_i to provide a zero-knowledge proof of knowledge (ZKPoK) of the discrete log of y_i/y_{i-1} in base g.

References

[1] M. Burmester, Yvo Desmedt, Hiroshi Doi, Masahiro Mambo, Eiji Okamoto, Mitsure Tada, and Y. Yoshifuji, *a structured ElGamal-Type multisignature scheme*, Advances in Cryptology-Proceedings of PKC'2000, Lecture notes in computer science,(2000), Springer-Verlag, 466-482.

CRYPTANALYSIS OF LKK PROXY SIGNATURE*

ZHENG Dong, LIU Shengli, CHEN Kefei
Department of Computer Science and Engineering
Shanghai Jiao Tong University, Shanghai 200030, P.R.China China
{zheng-dong,liu-sl,chen-kf}@cs.sjtu.edu.cn

Abstract   A strong proxy signature scheme [7] based on Schnorr's scheme was proposed by B. Lee *et al.* in 2001. In this paper we show that in the forementioned scheme, original signer may misuse a proxy signer's signature of a message M to forge the proxy signer's normal signature of M.

Keywords: cryptanalysis, digital signature, proxy signature

1.   Introduction

Digital signatures play a more and more important role in distributed environment. With digital signature[1,2,3], the transmissions of messages on Internet can achieve authenticity, data-integrity, and non-repudiation. The traditional handwriting are replacing by digital ones. Digital signature schemes can provide the cryptographic services: authentication, data integrity, and non-repudiation. Sometimes, we have the following scenarios: a department manager, say A, is responsible for signing some documents. However, he is busy with other important business, and has no time to sign these documents or he is not in the office upon the time. In those cases, A would like to delegate his signing capability to his secretary, say B, so B would sign documents on behalf of A if A is not available. In the above scenario, we need a so-called proxy signature scheme: a potential signer A delegates his signing capability to a proxy signer, B (in some way, A tells B what kind of messages B can sign), and B signs a message on behalf of the original signer A. the recepient of the message verifies the signature of B and the delegation of A together. Since the concept of proxy signature was introduced by Mambo et al.[4] in 1996, many proxy signature schemes were proposed [4,5,6,7], all of which are based on Schnorr's signature scheme[3]. According to the undeniability property, the

*Partial funding provided by NSFC under grants 90104005, 60173032 , 60273049 and 60303026.

proxy signature schemes are classified into two models: strong proxy signature and weak proxy signature in [7].

- Strong proxy signature: it represents both original signer's and proxy signer's signatures. Once a proxy signer creates a valid proxy signature, he cannot repudiate his signature creation against anyone.

- Weak proxy signature: it represents only original signer's signature. It does not provide non-repudiation of proxy signer.

In [7], B.Lee, H. Kim, and K. Kim also proposed a strong proxy signature scheme, which we will call LKK scheme. In this paper, we will show that LKK scheme is vulnerable to a new attack. In Section II, the brief review of Schnorr's scheme and LKK strong proxy signature scheme are given. Then we describe our new attack against LKK scheme. Section III concludes this paper.

2.   Brief review of related schemes and our attack

2.1   Schnorr's scheme [3]

Let us first how Schnorr's digital signature scheme works.

Let p and q be larger primes with $q|p-1$. Let g be a generator of a multiplicative subgroup of Z_p^* with order q, $H(\cdot)$ denotes a collision resistant hash function.

A signer **A** has a private key $x_A \in Z_q^*$ and the corresponding public key $y_A = g^{x_A} \mod p$. To sign a message M, **A** acts as follows:

1 Choose a random $k \in Z_q^*$;

2 Compute $r = g^k \mod p$ and $s = k + x_A H(M, r) \mod q$;

3 Define the signature on M to be the pair (r, s).

The signature is verified by checking that

$$g^s = r y_A^{H(M,r)} \mod p. \tag{1}$$

2.2   LKK strong proxy signature scheme

The following proxy signature scheme has been introduced in [7]. It is based on the above schnorr's scheme.

Suppose that the original signer **A** has a key pair (x_A, y_A), with x_A **A**'s private key and $y_A = g^{x_A} \mod p$ his public key. The (future) proxy signer **B** also has his own key pair (x_B, y_B), with x_B private key and $y_B = g^{x_B} \mod p$ public key.

Generation of the proxy key. The original signer **A** uses Schnorr's scheme to sign warrant information M_ω, which specifies what kind of messages **A** will allow the proxy **B** to sign on his behalf.

More precisely, **A** chooses at random $k_A \in Z_q^*$, and computes $r_A = g^{k_A} \bmod p$ and $s_A = k_A + x_A H(M_\omega, r_A) \bmod q$, Signer **A** sends (M_ω, r_A, s_A) to proxy signer **B** secretly.

After **B** gets (M_ω, r_A, s_A), he verifies the validity of the Schnorr's signature by checking whether the following equation holds:

$$g^{s_A} = r_A y_A^{H(M_\omega, r_A)} \quad \bmod p. \tag{2}$$

If eq.(2) holds, **B** computes his proxy key pair (x_P, y_P) in this way: the private proxy key is

$$x_P = x_B + s_A, \tag{3}$$

and the public proxy key is

$$y_P = g^{x_P} (= y_B r_A y_A^{H(M_\omega, r_A)}) \quad \bmod p. \tag{4}$$

Proxy signature generation. In order to create a proxy signature on a message M conforming to the warrant information M_ω, proxy signer **B** uses Schnorr's signature scheme with keys (x_P, y_P) and obtains a signature (r_P, s_P) for the message M. The valid proxy signature will be the tuple $(M, r_P, s_P, M_\omega, r_A)$.

Verification. A recipient can verify the validity of the proxy signature by checking that M conforms to M_ω and the verification equality of Schnorr's signature scheme with public key $y_P (= y_B r_A y_A^{H(M_\omega, r_A)}) \bmod p$.

Accept the proxy signature if and only if

$$g^{s_P} = r_P (y_B r_A y_A^{H(M_\omega, r_A)})^{H(M, r_P)} \tag{5}$$

holds.

The authors claimed that the scheme satisfies the following security requirements [7]: strong unforgeability, verifiability, strong identifiability, strong undeniability and prevention of misuse. In next section, we will present a new attack on LKK scheme.

3. Our attack

If the original signer A is dishonest, he can forge the signature of B on message M from a proxy signature.

After obtain the proxy signature $(M, r_P, s_P, M_\omega, r_A)$, the original signer A may forge B's signature on message M as follows:

1 compute $s' = x_A H(M, r_P) \bmod q$;

2 compute $s_B = s_P - s' \mod q$, and take $r_B = r_P$.

Then (r_B, s_B) and M satisfy eq. (1), i.e. $g^{s_B} = r_B y_B^{H(M, r_B)} \mod p$.
Suppose that

$$r_B = r_P = g^{k_P} \mod p, \; s_P = k_P + x_P H(M, r_P) \mod q,$$

where k_P is the random number selected by **B** for proxy signature on M. Then

$$s_B = s_P - s' = k_P + x_B H(M, r_B) \mod q$$

it is obviously that (r_B, s_B) is **B**'s Schonrr signature for message M.

In other words, (M, r_B, s_B) is the forged **B**'s signature on message M.

Remark. J. Herranz et al.[8] claim that other signature schemes (ElGamal signature or DSS) can be used in LKK strong proxy signature scheme. It should be noted that our attack works as well if DSS is used.

4.   Summary

Lee et al. briefly modified the proposal of [5] and get a strong proxy signature scheme (LKK scheme)[7]. However, the strong proxy signature scheme has a security flaw. We showed in this paper that in LKK scheme, the original signer **A** is able to misuse his power to forge a proxy signer **B**'s signature for a message, which has been signed by **B** as a proxy signature. Due to the attack, the original signer may confuse his responsibility with the proxy signer's.

References

[1] T.ElGamal, *public key cryptosystem and a signature scheme based on discrete* . IEEE Trans. IT-31, 1985

[2] L.Harn, *New digitral signature scheme based on discrete logarithm*. Electron. Lett., vol.30, no. 5, 1994.

[3] C.P. Schnorr, *Efficient signature generation by smart cards.* J of Cryptology," vol.4, 1991.

[4] M.Mambo, K.Usuda, and E.Okamoto, *Proxy signatures: Delegation of the power to sign messages.* IEICE Trans. E79-A, (9), 1996.

[5] S.Kim, S.Park, and D.Won, *Proxy signatures, revisited.* Proc. ICICS'97, LNCS 1334, 1997.

[6] W B Lee, C Y Chang, *Efficient proxy-protected proxy signature scheme based on discrete logarithm*, Proc. of 10th Conference on Information Security, Hualien, Taiwan, 2000.

[7] B.Lee, H. Kim, and K. Kim. *Strong proxy signature and its applications*, SCIS, 2001.

ATTACK ON IDENTITY-BASED BROADCASTING ENCRYPTION SCHEMES*

Shengli Liu, Zheng Dong, Kefei Chen

Department of Computer Science and Engineering
Shanghai Jiao Tong University, Shanghai 200030, P.R.China
{liu-sl, zheng-dong, chen-kf}@cs.sjtu.edu.cn

Abstract   The concept of "identity-based broadcasting encryption" (IBBE) was presented and two IBBE schemes were proposed recently by Y. Mu *et.al.* in [1]. Here we show that the two IBBE schemes are suffering a linear attack. We also point out there is a wrong assumption in their schemes.

Keywords: attack, broadcasting, encryption, pairing

1.   Introduction

Broadcast encryption systems are now more and more popular. A typical application of broadcast encryption systems is the broadcast of a pay TV programm. The broadcast center encrypts the TV programm with a session key, and broadcasts the encrypted signals to users. Only those who paid for the programm are able to decrypt the encrypted TV programm. An important issue in such a system is how to distribute the session key to the paid users. With the session key, those users are able to decrypt the broadcast. One way to solve the problem is that each user will be equipped with a decoding box. The broadcast center will first determine the group of users who paid for the programm. It then decides a session key, encrypts the session key with a public key algorithm, and broadcasts the encrypted session key to all users. Only the group of users who paid is able decrypt the encrypted session key with their decoding boxes. After getting the session key, the decoding boxes will decrypt the signals of the encrypted TV broadcast programm for users.

In [1], Y. Mu *et.al.* proposed to use an identity-based public key system to solve the problem. In an identity-based public key system, each user's public key can be derived from his identity information, and his corresponding private

*The project is supported by NSFC under grants 60303026, 90104005, 60173032.

key is determined by his public key. Each user in the broadcast encryption system registers his own identity information and gets corresponding private key from the broadcast center. The broadcast center determines the group of users who paid for the programm. The center collects all the identity information of the group to determine the public key of the group. The center then chooses a session key, and use the public key of the group to encrypt the session key. The center broadcasts the encrypted version of session key to all users. Only the members of the paid group are capable of decrypting the encrypted session key with their own private keys.

In the next two section, we will introduce two IBBE schemes proposed by Y. Mu *et.al*, and give a linear attack on them. In Section 4, we also point out a wrong assumption for the IBBE schemes. Section 5 concludes this paper.

2. Identity-Based Broadcasting Scheme: MSL Scheme 1

In the rest of the paper, we will call the IBBE schemes proposed by Y. Mu*et.al* MSL Schemes. MSL schemes are based on bilinear pairing. Before presenting MSL schemes, let us first see what are bilinear pairings.

2.1 Bilinear pairings

Let G_1 be a cyclic additive group generated by P, whose order is a prime q, and G_2 be a cyclic multiplicative group of the same order q. We assume that the discrete logarithm problems (DLP) in both G_1 and G_2 are hard. Let $e : G_1 \times G_1 \to G_2$ be a pairing which satisfies the following conditions:

1 Bilinear: $e(P_1 + P_2, Q) = e(P_1, Q)e(P_2, Q)$ and $e(P, Q_1 + Q_2) = e(P, Q_1)e(P, Q_2)$;

2 Non-degenerate: There exists $P \in G_1$ and $Q \in G_1$ such that $e(P, Q) \neq 1$;

3 Computability: There is an efficient algorithm to compute $e(P, Q)$ for all $P, Q \in G_1$.'

We note that the Weil and Tate pairings associated with supersingular elliptic curves or abelian varieties can be modified to create such bilinear maps. We refer to [2, 3, 4, 5, 6] for more details.

2.2 MSL Scheme 1

Now let us see the details of the first MSL in [1].

Suppose that in a TV broadcast system, there are totally n users. We denote the n users by $U_1, U_2, \ldots, U_n$. The broadcast center provides different kinds of programs to users. Those users who paid for a specific programm form a group. We denote different user groups by Group ID_i, $i = 1, 2, \ldots,$ where ID_i is the identity information of some group.

The broadcast center will choose and pre-compute some parameters as follows.

1 The center chooses public parameters (G_1, G_2, e, H_2):

- an additive group G_1 of order q.
- a multiplicative group G_2 of the same order q.
- a bilinear mapping $e(\cdot, \cdot) : G_1 \times G_1 \to G_2$.
- a strong hash function $H_2 : G_2 \to \{0, 1\}^l$.

2 The center determines the private parameters $(x, P, u_i', u_i, q, v, v_i, H_1), i = 1, 2, \ldots, n$ as follows:

- choose a prime $x \in \mathbb{Z}_q$ as its master key, and $P \in G_1$.
- choose a prime p_i and an integer k_i for user U_i, $i = 1, 2, \ldots, n$. Compute $u_i' = p_i^{k_i}$ and $u_i = \prod_{j=1, j \neq i}^{w} p_j^{k_j} + 1$. According to [1], the doublet (u_i, u_i') is a "qualified pair" associated with user U_i.
- let $q = \prod_{i=1}^{n} p_i^{k_i}$.
- choose a prime number v such that $gcd(v, q) = 1$.
- compute v_i such that $v \cdot v_i \mod q = u_i$.
- let $H_1 : \{0, 1\}^m \to G_1$ be a strong hash function.

Now there is a user group, whose identity is $ID_l \in \{0, 1\}^m$, consists of users $U_{l_1}, U_{l_2}, \ldots, U_{l_w}$. The center determines the decryption key D_{l_i} for user U_{l_i}, $i = 1, 2, \ldots, w$, in the group as follows.

1 Compute (d_{l_i}, u, Q_{ID_l}) and keep them secret.

- Set $d_{l_i} \leftarrow (x u_{l_i} + 1) v_{l_i} \mod q$.
- Extract Q_{ID_l} from the group identifier ID_l: $Q_{ID_l} \leftarrow H_1(ID_l)$.

2 Distribute the decryption key $D_{l_i} \leftarrow d_{l_i} Q_{ID_l}$ to user U_{l_i}.

The center chooses a session key K_s, and distributes K_s to Group ID_l with the following broadcast encryption scheme.

Encryption key (E_1, E_2): The center determines encryption key (E_1, E_2)

- Set $u \leftarrow \prod_{i=1}^{w} u_{l_i}' \mod q$.
- Let $E_1 = uP$ and $E_2 = uvP$.

Encryption: With (E_1, E_2, x), the center will encrypt K_s in the following way.

- Choose a random number $r \in \mathbb{Z}_q$. Let $R = rE_2$.
- Let $b = e(E_1, (x+1)Q_{ID})$, and $c = K_s \oplus H_2(b^r)$.
- The ciphertext for K_s is (R, c).

Decryption: User U_{l_i} decrypts the ciphertext (R, c) with his own decryption key D_{l_i} as follows.

- Compute

$$
\begin{aligned}
e(R, D_{l_i}) &= e(ruvP, (xu_{l_i} + 1)v_{l_i}Q_{ID_l}) \\
&= e(ruP, (xu_{l_i} + 1)u_{l_i}Q_{ID_l}) \\
&= e(rE_1, (x+1)Q_{ID_l}) = b^r
\end{aligned}
\tag{1}
$$

- Recover the session key $K_s = c \oplus H_2(b^r)$.

The center will encrypt the TV programm with K_s and broadcast the encrypted programm to every user. But only members of Group ID_l are able to watch the programm through decryption.

2.3   Linear Attack on MSL Scheme 1

MSL scheme 1 has an outstanding feature, which is that the center can dynamically add a new user or remove an existing user from a group without involvement of users. What the center should do just update the values of encryption keys (E_1, E_2).

Suppose that there is a group ID_l consisting of users $U_{l_1}, U_{l_2}, \ldots, U_{l_w}$.

- When adding a new user $U_{l_{w+1}}$, the center updates $E_1 \leftarrow U_{l_{w+1}}E_1$ and $E_2 \leftarrow U_{l_{w+1}}E_2$,

- When removing an existing user U_{l_j}, the center updates $E_1 \leftarrow U_{l_j}^{-1}E_1$ and $E_2 \leftarrow U_{l_j}^{-1}E_2$.

We know that in an IBBE system, the important thing is to avoid a user to receive unpaid TV programs. A legal user U_i (the user who paid for the program) was issued a decryption key D_i. User U_i can give away his decryption key D_i to other illegal users. However, the broadcast center can trace the traitor U_i with the decryption key D_i.

However, we find that it is possible for a number of legal users to collude and construct a new decryption key from their decryption keys.

Suppose that in Group ID_l, there are t colluders: $U_{c_1}, U_{c_2}, \ldots, U_{c_t}$. Each colluder U_{c_j} has his own decryption key D_{c_j}, $j = 1, 2, \ldots, t$.

1 The t colluders choose integers $a_1, a_2, \ldots, a_t$ satisfying $a_1 + a_2 + \cdots + a_t = 1$.

2 Colluder U_{c_j} computes $a_j D_{c_j}$, $j = 1, 2, \ldots, t$.

3 The t colluders determines the value of $D = \sum_{j=1}^{t} a_j D_{c_j}$.

Now D is a forged decryption key. It is obvious that D is different from the decryption keys of the colluders. But with D, an illegal user is able to decrypt (R, c) to recover the session key K_s.

Decryption: An illegal user decrypts the ciphertext (R, c) with the forged decryption key D as follows.

- Compute

$$e(R, D) = e(R, \sum_{j=1}^{t} a_j D_{c_j}) = \prod_{j=1}^{t} e(R, D_{c_j})^{a_j}$$

From Eq. (1), we know that $e(R, D_{c_j}) = b^r$, therefore

$$e(R, D) = \prod_{j=1}^{t} (b^r)^{a_j} = (b^r)^{\sum_{j=1}^{t} a_j} = b^r.$$

- Recover the session key $K_s = c \oplus H_2(b^r)$.

Since the forged decryption key D is a linear combination of colluders' decryption keys, we call the attack "linear attack".

3. MSL Scheme 2 and Its Analysis

3.1 MSL Scheme 2

MSL Scheme 1 showed how the broadcast center broadcasts a session key K_s to a group secretly. Now suppose that the center is going to broadcast the session key to several groups, say Group ID_1, Group $ID_2, \ldots$, Group ID_k. With MSL Scheme 1, the center has to encrypt K_s with different encryption key triplet $(E_1^{ID_i}, E_2^{ID_i}, x)$ and sent the corresponding ciphertext (R_{ID_i}, c_{ID_i}) to Group ID_i, here $i = 1, 2, \ldots, k$. That means the center has to determine k encryption keys, which is proportional to the number of groups.

In [1], another scheme, which we call MSL Scheme 2, is proposed to deal with the problem of broadcasting messages to multiple groups. In MSL Scheme 2, the center only has to determine one encryption key and broadcast some ciphertexts to multiple groups. Below we will describe MSL Scheme 2, and show that the scheme is suffering from the linear attack as well.

Suppose that the session key K_s will be distributed to Group ID_1, Group $ID_2, \ldots,$ Group ID_k. In Group $ID_i, i = 1, 2, \ldots, k$, the members of users are denoted by $U_{i_1}^{ID_i}, U_{i_2}^{ID_i}, \ldots, U_{i_{l_i}}^{ID_i}$.

The center chooses three additional hash functions: $H_3 : \{0,1\}^l \times \{0,1\}^l \to \mathbb{Z}_q$, $H_4 : G_2 \to G_2$, and $H_5 : G_2 \to \{0,1\}^l$. Keep H_3, H_5 secret, and publish H_4.

The center determines encryption key (E_1, E_2).

- Set $u \leftarrow \prod_{i=1}^{k} \prod_{j=1}^{l_i} u_{i_j}'^{ID_i} \mod q$.
- Let $E_1 = uP$ and $E_2 = uvP$.

With (E_1, E_2, x), the center will encrypt K_s in the following way.

Encryption:
- Choose a random number $\sigma \in \mathbb{Z}_q$. Set $r \leftarrow H_3(\sigma, K_s)$. Let $R = rE_2$.
- Let $b_{ID_i} = e(E_1, (x+1)Q_{ID_i})$. Choose $b_{ID_i}' \in G_2$, and get $\hat{b}_{ID_i} = b_{ID_i} b_{ID_i}'$.
- Compute $c_{ID_i} \leftarrow b_{ID_i}'^r H_4(b_{ID_i}^r)$ and $c_{ID_i}' \leftarrow K_s \oplus H_2(\hat{b}_{ID_i}^r)$.
- The ciphertext of K_s for Group ID_i is (c_{ID_i}, c_{ID_i}', R).
- Broadcast $(c_{ID_1}, c_{ID_2}', \ldots, c_{ID_k}, c_{ID_k}', R)$ to the groups.

Decryption: User $U_{i_j}^{ID_i}, j = 1, 2, \ldots, l_i$, decrypts the ciphertext (c_{ID_i}, c_{ID_i}', R) with his own decryption key $D_{i_j}^{ID_i}$ as follows.

- Compute

$$\begin{aligned}
e(R, D_{i_j}^{ID_i}) &= e(ruvP, (xu_{i_j}+1)v_{i_j}Q_{ID_i}) \\
&= e(ruP, (xu_{i_j}+1)u_{i_j}Q_{ID_i}) \\
&= e(rE_1, (x+1)Q_{ID_i}) = b_{ID_i}^r \qquad (2)
\end{aligned}$$

- Compute $b_{ID_i}'^r = c_{ID_i} \cdot \left(H_4(b_{ID_i}^r)\right)^{-1}$, and $\hat{b}_{ID_i} = b_{ID_i} \cdot b_{ID_i}'$.
- Recover the session key $K_s = c_{ID_i}' \oplus H_2(\hat{b}_{ID_i}^r)$.

3.2   Linear Attack on MSL Scheme 2

Linear attack works on MSL Scheme 2 in the same way. Suppose that in some group, say Group ID_k, there are t colluders: $U_{c_1}^{ID_k}, U_{c_2}^{ID_k}, \ldots, U_{c_t}^{ID_k}$. Each colluder $U_{c_j}^{ID_k}$ has his own decryption key $D_{c_j}^{ID_k}, j = 1, 2, \ldots, t$. The attack is similar to the previous one on MSL Scheme 1.

1 The t colluders choose integers $a_1, a_2, \ldots, a_t$ satisfying $a_1 + a_2 + \cdots + a_t = 1$.

2 Colluder $U_{c_j}^{ID_k}$ computes $a_j D_{c_j}^{ID_k}$, $j = 1, 2, \ldots, t$.

3 The t colluders determines the value of $D = \sum_{j=1}^{t} a_j D_{c_j}^{ID_k}$.

The session key K_s can be recovered with the forged decryption key D in the following way.

Decryption: An illegal user decrypts the ciphertext (c_{ID_k}, c'_{ID_k}, R) with the forged decryption key D as follows.

- Compute

$$
e(R, D) = e\left(R, \sum_{j=1}^{t} a_j D_{c_j}^{ID_k}\right) = \prod_{j=1}^{t} e\left(R, D_{c_j}^{ID_k}\right)^{a_j}
$$

$$
= \left(b_{ID_k}^r\right)^{\sum_{j=1}^{t} a_j} = b_{ID_k}^r
$$

- Compute $b_{ID_k}'^r = c_{ID_k} \cdot \left(H_4(b_{ID_k}^r)\right)^{-1}$, and $\hat{b}_{ID_k} = b_{ID_k} \cdot b'_{ID_k}$.

- Recover the session key $K_s = c'_{ID_k} \oplus H_2(\hat{b}_{ID_k}^r)$.

Given a forged decryption key $D = \sum_{j=1}^{t} a_j D_{c_j}$, if the broadcast center can uniquely determine the value of $D_{c_1}, D_{c_2}, \ldots, D_{c_t}$, then the identity of traitors can be traced. However, in our linear attack, there may exist another set of user $U_{d_1}, U_{d_2}, \ldots, U_{d_s}$ and integers $b_1, b_2, \ldots, b_s$, such that $D = \sum_{j=1}^{t} a_j D_{c_j} = \sum_{i=1}^{s} b_i D_{d_i}$. Therefore, the forged decryption key is untraceable to any of the traitors.

4.   Remark on the Assumption of the Order of the Group

Bilinear pairing, including the Weil pairing and the Tate pairing, is an essential tool to construct identity-based cryptosystems. In [1], the authors suggested to use the Weil pairing for their identity-based encryption schemes (MSL schemes). Suppose that E is an supersingular elliptic curve over a field K with a positive characteristic, and $E[q]$ denotes a q-torsion group of E. The Weil pairing $e(\cdot, \cdot)$ is a mapping from $E[q] \times E[q]$ to a multiplicative group G_2 of order q of some extension field of K. Here, $E[q]$ functions as the additive group G_1 of order q.

It should be noted that to compute the Weil pairing, the parameter q must be known (see [2, 3, 4]).

In [1], however, the order of the additive group is only known by the broadcast center. No user knows the value of q. But the members of groups are required to compute some Weil pairing for decryption. Without q, it is impossible for the members to implement decryption.

If q is public to all users, as [1] pointed out, there will be a security threat. We now consider MSL Scheme 1. The t colluders $U_{c_1}, U_{c_2}, \ldots, U_{c_t}$, can just simply add their decryption keys and get $D = D_{c_1} + D_{c_2} + \cdots + D_{c_t}$. Since $e(R, D_{c_i}) = b^r, i = 1, 2, \ldots, t,$

$$e(R, D) = e(R, D_{c_1})e(R, D_{c_2}) \ldots e(R, D_{c_t}) = b^{rt}.$$

When $gcd(t, q) = 1$, $t^{-1} \bmod q$ can be easily computed with Euclid's extension algorithm, so $e(R, D)^{t^{-1}}$, i.e. b^r, can be easily determined. After getting b^r, the session key K_s results directly from $K_s = c \oplus H_2(b^r)$.

The same problem exists in MSL Scheme 2.

5.   Conclusion

In this paper, we presented a linear attack on the identity-based broadcast encryption schemes proposed by Y. Mu *et.al*. The linear attack showed the legal users in the same group can collude and forge a new decryption key with their own decryption keys. In the mean time, with the linear attack, it is impossible for trace the identity of the colluders. On the other hand, we point out that the assumption for MSL schemes are not rational. The reason is that users are not able to do decryption if the order q of the group G_1 (and G_2) is unknown to users. On the other hand, if q is known by users, users can successfully collude.

References

[1] Y. Mu, W. Susilo, Y. Lin, *identity-based broadcasting*. Indocrypt2003, LNCS 2904, pp.177-190, Springer-Verlag, 2003.

[2] D. Boneh and M. Franklin, *Identity-based encryption from the Weil pairing*, Advances in Cryptology-Crypto'2001, LNCS 2139, pp. 213-229, Springer-Verlag, 2001.

[3] D. Boneh, B. Lynn, and H. Shacham, *Short signatures from the Weil pairing*, In C. Boyd, editor, Advances in Cryptology-Asciacrypt'2001, LNCS 2248, pp. 514-532, Springer-Verlag, 2001.

[4] S. D. Galbraith, K. Harrison, and D. Soldera, *Implementing the Tate pairing*, In C. Fieker and D.R. Kohel (Eds.): ANTS 2002, LNCS 2369, pp. 324-337, Springer-Verlag, 2002.

[5] P.S.L.M. Barreto, H.Y.Kim, B.Lynn, and M.Scott, *Efficient algorithms for pairing-based cryptosystems*, In Advances in Cryptology, Crypto2002, LNCS 2442, pp.352-368, Springer-Verlag, 2002.

[6] F. Zhang and K. Kim, *ID-based blind signature and ring signature from pairings*, In Proc. of Asciacrypt2002, LNCS 2501, pp. 533-547, Springer-Verlag, 2002

DIFFERENTIAL-LINEAR CRYPTANALYSIS OF CAMELLIA[*]

Wenling WU, Dengguo FENG
State Key Laboratory of Information Security
Institute of Software
Chinese Academy of Sciences
Beijing 100080, PR CHINA
{wwl,fdg}@is.iscas.ac.cn

Abstract   Camellia is the final selection of 128-bit block cipher in NESSIE. In this paper, we present differential-linear cryptanalysis of modified camellia reduced to 9 and 10 rounds. For modified camellia with 9 rounds we can find the user key with 2^{14} chosen plaintexts and $2^{185.5}$ encryptions and for modified camellia with 10 rounds we can find the user key with 2^{14} chosen plaintexts and $2^{245.6}$ encryptions.

Keywords: block cipher, differential-linear cryptanalysis, data complexity, time complexity

1.   Introduction

Camellia [1] is a 128-bit block cipher which was published by NTT and Mitsubishi in 2000 and recently selected as the final selection of the NESSIE [2] project, and also suggested as a candidate for the CRYPTREC project in Japan [3]. The security of Camellia has been studied by many researchers [4 $\sim$ 12]. The security of Camellia against higher-order differential cryptanalysis is discussed in [4] and [5]. A truncated differential attack on 8-round variant of Camellia without FL/FL^{-1} functions is presented in [6] requiring $2^{55.6}$ encryptions and $2^{83.6}$ chosen plaintexts. Truncated and impossible differential cryptanalysis of Camellia without FL/FL^{-1} functions is described in [7]. A differential attack on 9 rounds Camellia without FL/FL^{-1} functions is proposed in [8]£¬requiring 2^{105} chosen plaintexts. The security of Camellia against Square attack is discussed in [9] and [10]. Yeom et.al. have studied

[*] This work was supported by Chinese Natural Science Foundation (Grant No. 60373047)and 863 Project (Grant No. 2003AA14403)

integral properties and apply them to Camellia in [11].Furthermore,collision attack on reduced-round camellia is introduced in [12].

In this paper we present differential-linear cryptanalysis on modified Camellia reduced to 9 and 10 rounds. Section 2 briefly describes the structure of Camellia, A 4-round differential characteristic with probability 1 is explained in Section 3. In Section 4, we show how to use the 4-round differential characteristic and a 1-round linear approximation to attack on modified Camellia reduced to 9 and 10 rounds.Finally, in Section 5 we summarize this paper.

2.   Description of the Camellia

Camellia has a 128 bit block size and supports 128,192 and 256 bit keys. The design of Camellia is based on the Feistel structure and its number of rounds is 18(128 bit key) or 24(192/256 bit key). The FL/FL^{-1} function layer is inserted at every 6 rounds. Before the first round and after the last round, there are pre- and post-whitening layers which use bitwise exclusive-or operations with 128 bit subkeys, respectively. But we will consider camellia without FL/FL^{-1} function layer and whitening layers and call it modified camellia.

Let L_{r-1} and R_{r-1} be the left and the right halves of the rth round inputs, and k_r be the rth round subkey. Then the Feistel structure of Camellia can be written as

$$L_r = R_{r-1} \oplus F(L_{r-1}, k_r),$$
$$R_r = L_{r-1},$$

here F is the round function defined below:

$$F : F_2^{64} \times F_2^{64} \longrightarrow F_2^{64},$$
$$(X_{64}, k_{64}) \longrightarrow Y_{(64)} = P(S(X_{(64)} \oplus k_{(64)})).$$

where S and P are defined as follows:

$$S : F_2^{64} \longrightarrow F_2^{64},$$

$$l_{1(8)}||l_{2(8)}||l_{3(8)}||l_{4(8)}||l_{5(8)}||l_{6(8)}||l_{7(8)}||l_{8(8)}$$
$$\longrightarrow l_{1(8)}^*||l_{2(8)}^*||l_{3(8)}^*||l_{4(8)}^*||l_{5(8)}^*||l_{6(8)}^*||l_{7(8)}^*||l_{8(8)}^*$$

$$\begin{aligned}
l_{1(8)}^* &= s_1(l_{1(8)}), & l_{5(8)}^* &= s_2(l_{5(8)}), \\
l_{2(8)}^* &= s_2(l_{2(8)}), & l_{6(8)}^* &= s_3(l_{6(8)}), \\
l_{3(8)}^* &= s_3(l_{3(8)}), & l_{7(8)}^* &= s_4(l_{7(8)}), \\
l_{4(8)}^* &= s_4(l_{4(8)}), & l_{8(8)}^* &= s_1(l_{8(8)}).
\end{aligned}$$

$$P : F_2^{64} \longrightarrow F_2^{64},$$

$$Z_{1(8)}||Z_{2(8)}||Z_{3(8)}||Z_{4(8)}||Z_{5(8)}||Z_{6(8)}||Z_{7(8)}||Z_{8(8)}$$
$$\longrightarrow Z_{1(8)}^*||Z_{2(8)}^*||z_{3(8)}^*||Z_{4(8)}^*||Z_{5(8)}^*||Z_{6(8)}^*||Z_{7(8)}^*||Z_{8(8)}^*$$

$$Z_1^* = Z_1 \oplus Z_3 \oplus Z_4 \oplus Z_6 \oplus Z_7 \oplus Z_8, \quad Z_5^* = Z_1 \oplus Z_2 \oplus Z_6 \oplus Z_7 \oplus Z_8,$$
$$Z_2^* = Z_1 \oplus Z_2 \oplus Z_4 \oplus Z_5 \oplus Z_7 \oplus Z_8, \quad Z_6^* = Z_2 \oplus Z_3 \oplus Z_5 \oplus Z_7 \oplus Z_8,$$
$$Z_3^* = Z_1 \oplus Z_2 \oplus Z_3 \oplus Z_5 \oplus Z_6 \oplus Z_8, \quad Z_7^* = Z_3 \oplus Z_4 \oplus Z_5 \oplus Z_6 \oplus Z_8,$$
$$Z_4^* = Z_2 \oplus Z_3 \oplus Z_4 \oplus Z_5 \oplus Z_6 \oplus Z_7, \quad Z_8^* = Z_1 \oplus Z_4 \oplus Z_5 \oplus Z_6 \oplus Z_7.$$

Below briefly describes the key schedule of Camellia. First two 128-bit variables K_L and K_R are generated from the user key. Then two 128-bit variables K_A and K_B are generated from K_L and K_R. Note that K_B is used only when the user key is of 192 or 256 bits. The round subkeys are generated by rotating K_L, K_R, K_A and K_B. Details are shown in [1].

3.   4-Round Distinguisher

Choose two plaintexts $P = (L_0, R_0)$ and $P^* = (L_0^*, R_0^*)$:

$$L_0 = (\alpha_1, \alpha_2, \cdots, \alpha_8), \qquad R_0 = (x, \beta_2, \cdots, \beta_8).$$

$$L_0^* = L_0, \qquad R_0^* = (x^*, \beta_2, \cdots, \beta_8).$$

where $x \neq x^*$, α_i and β_j are constants in F_2^8. Thus, the input of the 2nd round can be written as follows:

$$L_1 = (x \oplus \gamma_1, \gamma_2, \cdots, \gamma_8), \qquad R_1 = (\alpha_1, \alpha_2, \cdots, \alpha_8),$$

$$L_1^* = (x^* \oplus \gamma_1, \gamma_2, \cdots, \gamma_8), \qquad R_1^* = (\alpha_1, \alpha_2, \cdots, \alpha_8),$$

where γ_i are entirely determined by $\alpha_i (1 \leq i \leq 8), \beta_j (2 \leq j \leq 8)$ and k_1, so γ_i are constants when the user key is fixed. In the 2nd round a transformation on L_1 and L_1^* using $F(\bullet, k_2)$ is as follows:

$$
\begin{aligned}
L_1 \;=\; & (x \oplus \gamma_1, \gamma_2, \cdots, \gamma_8) \xrightarrow{F(\bullet, k_2)} \\
& (y \oplus \theta_1, y \oplus \theta_2, y \oplus \theta_3, \theta_4, y \oplus \theta_5, \theta_6, \theta_7, y \oplus \theta_8) \\
L_1^* \;=\; & (x^* \oplus \gamma_1, \gamma_2, \cdots, \gamma_8) \xrightarrow{F(\bullet, k_2)} \\
& (y^* \oplus \theta_1, y^* \oplus \theta_2, y^* \oplus \theta_3, \theta_4, y^* \oplus \theta_5, \theta_6, \theta_7, y^* \oplus \theta_8)
\end{aligned}
$$

where $y = s_1(x \oplus \gamma_1 \oplus k_{2,1})$, $y^* = s_1(x^* \oplus \gamma_1 \oplus k_{2,1})$, $k_{2,1}$ is the first byte of k_2, θ_i are entirely determined by $\gamma_i (1 \leq i \leq 8)$ and k_2, thus θ_i are constants when the user key is fixed. Therefore, the output of the 2nd round is

$$
\begin{aligned}
L_2 &= (y \oplus \varpi_1, y \oplus \varpi_2, y \oplus \varpi_3, \varpi_4, y \oplus \varpi_5, \varpi_6, \varpi_7, y \oplus \varpi_8), \\
R_2 &= L_1 = (x \oplus \gamma_1, \gamma_2, \cdots, \gamma_8), \\
L_2^* &= (y^* \oplus \varpi_1, y^* \oplus \varpi_2, y^* \oplus \varpi_3, \varpi_4, y^* \oplus \varpi_5, \varpi_6, \varpi_7, y^* \oplus \varpi_8), \\
R_2^* &= L_1^* = (x^* \oplus \gamma_1, \gamma_2, \cdots, \gamma_8).
\end{aligned}
$$

where $\varpi_i = \theta_i \oplus \alpha_i$ are constants. In the 3rd round a transformation on L_2 and L_2^* using $F(\bullet, k_3)$ is as follows:

$$L_2 = (y \oplus \varpi_1, y \oplus \varpi_2, y \oplus \varpi_3, \varpi_4, y \oplus \varpi_5, \varpi_6, \varpi_7, y \oplus \varpi_8)$$

$$\xrightarrow{F(\bullet, k_3)} (z_1, z_2, \cdots, z_8)$$

$$L_2^* = (y^* \oplus \varpi_1, y^* \oplus \varpi_2, y^* \oplus \varpi_3, \varpi_4, y^* \oplus \varpi_5, \varpi_6, \varpi_7, y^* \oplus \varpi_8)$$

$$\xrightarrow{F(\bullet, k_3)} (z_1^*, z_2^*, \cdots, z_8^*)$$

By observing the round function we find

$$z_3 \oplus z_4 \oplus z_5 \oplus z_6 \oplus z_7 = s_4(\varpi_7 \oplus k_{3,7}) \oplus \sigma_1$$
$$z_3^* \oplus z_4^* \oplus z_5^* \oplus z_6^* \oplus z_7^* = s_4(\varpi_7 \oplus k_{3,7}) \oplus \sigma_1$$

σ_1 is entirely determined by $\varpi_i(1 \leq i \leq 8)$ and k_3, so σ_i is a constant when the user key is fixed.

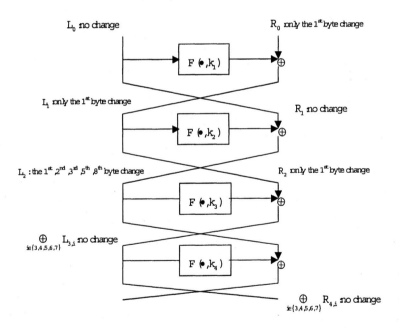

Figure 1.   4-rounds differential characteristic

Thus,we have the left half of output for the 3rd round:

$$L_3 = (z_1 \oplus x \oplus \gamma_1, z_2 \oplus \gamma_2, z_3 \oplus \gamma_3, \cdots, z_8 \oplus \gamma_8).$$

$$L_3^* = (z_1^* \oplus x^* \oplus \gamma_1, z_2^* \oplus \gamma_2, z_3^* \oplus \gamma_3, \cdots, z_8^* \oplus \gamma_8).$$

So the right half of output for the 4th round is as follows:

$$R_4 = L_3 = (z_1 \oplus x \oplus \gamma_1, z_2 \oplus \gamma_2, z_3 \oplus \gamma_3, \cdots, z_8 \oplus \gamma_8).$$

$$R_4* = L_3^* = (z_1^* \oplus x^* \oplus \gamma_1, z_2^* \oplus \gamma_2, z_3^* \oplus \gamma_3, \cdots, z_8^* \oplus \gamma_8).$$

so we have the following equation:

$$R_{4,3} \oplus R_{4,4} \oplus R_{4,5} \oplus R_{4,6} \oplus R_{4,7} = R_{4,3}^* \oplus R_{4,4}^* \oplus R_{4,5}^* \oplus R_{4,6}^* \oplus R_{4,7}^* \quad (1)$$

with probability 1.

4.   Attacks on Camellia Reduced to 9 and 10 Rounds

In this section, we describe differential-linear cryptanalysis of modified Camellia reduced to 9 rounds in detail.

By testing

$$s_4 : F_2^8 \to F_2^8 \qquad X \to s_4(X) = Y,$$

we get the following linear approximation of s_4-box:

$$X[4] = Y[4] \quad (2)$$

with probability $p = 1/2 + 3/2^6$, where $X[m]$ denotes the mth bit of X.

If known $k_{1,1}$, we can choose two plaintexts $P = (L_0, R_0)$ and $P^* = (L_0^*, R_0^*)$ such that (L_1, R_1) and (L_1^*, R_1^*) satisfy the input condition of the 4-round differential characteristic in section 3. Let $I = \{3, 4, 5, 6, 7\}$, we have the following equation:

$$\bigoplus_{i \in I} R_{5,i} = \bigoplus_{i \in I} R_{5,i}^* \quad (3)$$

Further observing the round function we have

$$\bigoplus_{i \in I} R_{5,i} = s_4(R_{6,7} \oplus k_{6,7}) \oplus \left(\bigoplus_{i \in I} L_{6,i} \right) \quad (4)$$

$$\bigoplus_{i \in I} R_{5,i}^* = s_4(R_{6,7}^* \oplus k_{6,7}) \oplus \left(\bigoplus_{i \in I} L_{6,i}^* \right) \quad (5)$$

By using equation(2) we get

$$\bigoplus_{i \in I} R_{5,i}[4] = R_{6,7}[4] \oplus k_{6,7}[4] \oplus \left(\bigoplus_{i \in I} L_{6,i}[4] \right) \quad (6)$$

$$\bigoplus_{i \in I} R_{5,i}^*[4] = R_{6,7}^*[4] \oplus k_{6,7}[4] \oplus \left(\bigoplus_{i \in I} L_{6,i}^*[4] \right) \quad (7)$$

Equation(6)and (7) hold with probability $p = 1/2 + 3/2^6$.

Because $\bigoplus_{i \in I} R_{5,i}[4] = \bigoplus_{i \in I} R_{5,i}^*[4]$, we get

$$R_{6,7}[4] \oplus (\bigoplus_{i \in I} L_{6,i}[4]) = R_{6,7}^*[4] \oplus (\bigoplus_{i \in I} L_{6,i}^*[4]) \tag{8}$$

Equation(8)holds with probability $p^2 + (1 - p)^2 = 1/2 + 9/2^{11}$. From the round transformation we have

$$\begin{aligned} R_{6,7} &= s_3(R_{7,3} \oplus k_{7,3}) \oplus s_4(R_{7,4} \oplus k_{7,4}) \oplus s_2(R_{7,5} \oplus k_{7,5}) \oplus s_3(R_{7,6} \\ &\oplus k_{7,6}) \oplus s_1(R_{7,8} \oplus k_{7,8}) \oplus L_{7,3} \oplus L_{7,4} \oplus L_{7,5} \oplus L_{7,6} \oplus L_{7,8} \end{aligned}$$

Therefore if known $k_{7,3}, k_{7,4}, k_{7,5}, k_{7,6}, k_{7,8}, k_8, k_9$, $R_{6,7}[4] \oplus (\bigoplus_{i \in I} L_{6,i}[4])$ and $R_{6,7}^*[4] \oplus (\bigoplus_{i \in I} L_{6,i}^*[4])$ can be obtained through decrypting the ciphertexts $C = (L_9, R_9)$ and $C = (L_9^*, R_9^*)$.

When the correct value of $(k_{7,3}, k_{7,4}, k_{7,5}, k_{7,6}, k_{7,8}, k_8, k_9)$ is used, we expect $R_{6,7}[4] \oplus (\bigoplus_{i \in I} L_{6,i}[4]) = R_{6,7}^*[4] \oplus (\bigoplus_{i \in I} L_{6,i}^*[4])$ holds with probability $1/2 + 9/2^{11}$; when an incorrect value is used, the produced data is more random and we expect the probability closer to 0.5.

Based on Ref[11] that approximately $2^{25}/81$ pairs of chosen plaintexts are needed. Now we introduce how to obtain the desired pairs of plaintexts. Fixed $\alpha_i (2 \leq i \leq 8)$ and $\beta_j (1 \leq j \leq 8)$, we choose the set of plaintexts:

$$\begin{aligned} \phi = \{P^{(t)} = (L_0^t, R_0^t) | R_0^t &= (y \oplus \beta_1, y \oplus \beta_2, y \oplus \beta_3, \beta_4, y \oplus \beta_5, \beta_6, \beta_7, \\ y \oplus \beta_8), L_0^t &= (x, \alpha_2, \cdots, \alpha_8), t = x + 2^8 y\} \end{aligned}$$

y take value all over F_2^8, x take 64 values from F_2^8, thus $|\phi| = 2^{14}$. For any $P^{(t)}, t = a + b \times 2^8$, we can choose 63 plaintexts $P^{(t^*)}$ from $\phi, t^* = a_1 + b_1 \times 2^8, a_1 \neq a, b_1 = s_1(a \oplus k_{1,1}) \oplus s_1(a_1 \oplus k_{1,1}) \oplus b$, such that $P^{(t)}$ and $P^{(t^*)}$ are the desired pair. Therefore, we can construct $2^{14} \times 63/2 > 2^{25}/81$ pairs of desired plaintexts from ϕ.

Algorithm

Step1, choose the set ϕ of plaintexts and corresponding set of ciphertexts is denoted as Ω.

Step2, For each possible value g of $(k_{7,3}, k_{7,4}, k_{7,5}, k_{7,6}, k_{7,8}, k_8, k_9)$, decrypt the ciphertxts in Ω and compute $R_{6,7}[4] \oplus (\bigoplus_{i \in I} L_{6,i}[4])$. Let N_g be the total number of pairs of plaintexts that satisfy $R_{6,7}[4] \oplus (\bigoplus_{i \in I} L_{6,i}[4]) = R_{6,7}^*[4] \oplus (\bigoplus_{i \in I} L_{6,i}^*[4])$.

Step 3, output the key candidate g corresponding to the maximum value of all N_g.

The main time complexity of attack is the step2, the time of computing each $R_{6,7}[4] \oplus (\bigoplus_{i \in I} L_{6,i}[4])$ is about the 3-round encryption, so the time complexities of attack is about $2^{14} \times 2^{168} \times 3/9 < 2^{180.5}$ encryptions.

Using above Algorithm we can attack modified Camellia reduced to 10 rounds. The difference is in step 2, compute $R_{6,7}'[4] \oplus (\bigoplus_{i \in I} L_{6,i}[4])$ from ciphertexts for each possible candidates of $(k_{7,3}, k_{7,4}, k_{7,5}, k_{7,6}, k_{7,8}, k_8, k_9, k_{10})$. The time of computing each $R_{6,7}'[4] \oplus (\bigoplus_{i \in I} L_{6,i}[4])$ is about the 4-round encryption, so the time complexity of attack is about $2^{14} \times 2^{232} \times 4/10 < 2^{245.6}$ encryptions.

5. Conclusion

In this paper, we studied differential-linear cryptanalysis for modified Camellia reduced to nine and ten rounds. For modified Camellia with 9 rounds we can find user key with 2^{14} chosen plaintexts and $2^{180.5}$ encryptions and for modified Camellia with 10 rounds we can find the user key with 2^{14} chosen plaintexts and $2^{245.6}$ encryptions. Up to 10 rounds, the differential-linear cryptanalysis is a faster way to attack Camellia than the brute force key search.

References

[1] K.Aoki,T.Ichikawa,M.Kanda,M.Matsui,S.Moriai,J.Nakajima and T.Tokita, Specification of Camellia-a 128-bit Block Cipher, in Proceedings of Selected Areas in Cryptography -SAC'2000, in Lecture Notes in Computer Science, pp. 183-191, Springer-Verlag,2000.

[2] http://www.cryptonessie.org

[3] CRYPTREC project, http://www.ipa.go.jp/security/enc/CRYPTREC/

[4] T.Kawabata, T.Kaneko, A study on higher order differential attack of Camellia, Proceedings of the 2nd NESSIE workshop,2001.

[5] Y.Hatano,H.Sekine, and T.Kaneko, Higher order differential attack of Camellia(II), in Proceedings of Selected Areas in Cryptography-SAC'02, Lecture Notes in Computer Science, pp.39-56,Springer-Verlag,2002.

[6] M.Sugita,K.Kobara, and H.Imai, Security of reduced version of the block cipher Camellia against truncated and impossible differential cryptanalysis, in Proceedings of Asiacrypt'01(C.Boyd,ed.),no.2248 in Lecture Notes in Computer Science, pp.193-207,Springer-Verlag,2001.

[7] T.Shirai,S.Kanamaru, and G.Abe, Improved upper bounds of differential and linear characteristic probability for Camellia, in Proceedings of Fast Software Encryption-FSE'02(J.Daemen and V.Rijmen,ed.),no.2365 in Lecture Notes in Computer Science,pp.128-142,Springer-Verlag,2002.

[8] He Ye-ping and Qing Si-han, Square attack on Reduced Camellia Cipher, ICICS'01 (Qing Si-han,ed,), no.2229 in Lecture Notes in Computer Science,pp.238-245, Springer-Verlag,2001.

[9] Y.Yeom, S. Park, and I. Kim, On the security of Camellia against the square attack, in Proceedings of Fast Software Encryption-FSE'02(J.Daemen and V.Rijmen, ed.),no.2365 in Lecture Notes in Computer Science,pp.89-99,Springer-Verlag,2002.

[10] Y.Yeom, I. Park, and I. Kim, A study of Integral type cryptanalysis on Camellia, The 2003 Symposium on Cryptography and Information Security-SCIS'03

[11] S.K. Langford and M.E. Hellman, Differential-Linear Cryptanalysis,in Proceedings of Crptology-CRYPTO'94,pp.17-26,Springer-Verlag,1994.

[12] Wen-Ling Wu and Deng-Guo Feng, Collision Attack on Reduced-Round Camellia, Cryptology eprint Archive Report 2003/135,http://eprint.iacr.org/2003/135

SECURITY ANALYSIS OF EV-DO SYSTEM

Zhu, Hong Ru

Bell-labs, Lucent China, No.30 Hai Dian Nan Lu, Beijing 100080, China
hongruzhu@lucent.com

Abstract   EV-DO security architecture is introduced in this paper. Following that, the authentication flows in RAN level and IP level, encryption and integrity are investigated herein. In addition, the paper gave an analysis of the EV-DO security in detail. At last, some enhancements suggestions are proposed.

Keywords: EV-DO, access authentication, CHAP, security analysis

1.   INTRODUCTION

With the high development of Internet and wireless information technology, the requirement for wireless data service is increasing drastically. Also the service is required to provide instance, versatility and high quality. But the current CDMA1X technology is far to meet these requirements. So the 3GPP2 association, which is mostly composed, of North America countries held up EV work group in early 2000.This group proposed the EV-DO [1] technology to meet the high data rate requirements. The standard is re-defined as HDPR (High Data Packet Rate) this year.

HRPD specification has been finished in the end of 2001. Compared to current CDMA1X technology, it has the following advantages:

- Air interface:1xEV-DO effectively resolved the data service transmission bottleneck problem in the air interface. Compete with the 153.6Kbps of CDMA2000 1X,1xEV-DO forward link peak rate can be up to 2.4576 Mbps/(Sector).

- Frequency parameters:1xEV-DO and IS-95/CDMA2000 1X have same RF characteristics, Chip rate, power requirement, coverage, then protects the current operator invest in maximum.

- Architecture:1xEV-DO is very flexible in the network construction. Standalone network can provide service for the users who only need the packet service; and combined network with IS-95/CDMA2000 1X can provide voice and high data rate services in the same time.

So basically HRPD technology is low cost, small risk, fast rate, flexible to construct network, easy to implement. Hence, we believe the EV-DO technology can absolutely provide very fast and efficient data service to the users.

But for wireless data service, security is playing very important role. EV-DO (HRPD) has been configured with a mature set of security scheme. This paper shall introduce and analyze its security architecture, authentication flows, and encryption, message integrity in depth.

2.   EV-DO Security Architecture

EV-DO mainly provides security protection in the air interface between AT (access terminal) and AN (Access network), including encryption and signaling integrity protection. The authentication of EV-DO user is performed between PDSN and AT or performed between AT and AN. Next section shall describe this deeply. The architecture of EV-DO is shown as below. A security layer is

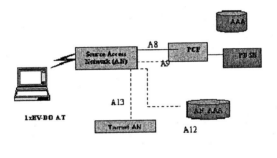

Figure 1.

included in the EV-DO air interface, placed between connection layer and MAC layer, which is made up of security parameters protocol, message authentication protocol, key agreement protocol and encryption protocol. Among them, security parameters protocol is to provide the parameters for encryption protocol, such as Time stamp, synchronization parameters. Authentication protocol refers to message authentication, i.e integrity protection. It has two options, DEFAULT no protection and SHA-1. Key agreement protocol is meant to provide key required for a session. Current standard defines two options, No or D-H. Encryption protocol defines no encryption or AES encryption.

3.   EV-DO User Authentication

EV-DO authentication is different from CDMA1X. It does not use the SSD,A-KEY,CAVE. The access authentication is performed in the PPP layer using CHAP[2] protocol, MD5 algorithm. Secret individual parameters (such as SS,

PASSWORD, NAI etc needed by CHAP) are required to store in the card in order to validate whether the user can be authorized the access of the wireless and IP network.

Actually, there are two kinds of authentications performed respectively in the air interface and the PPP session between PDSN and AT. One is called RAN Authentication, i.e inner CHAP, which is defined in IOS A.S0007; the other is called PDSN Authentication, also called by outer CHAP authentication, which is defined in IS-835.

Inner CHAP (RAN authentication) is optional, but outer CHAP between PDSN and AT is mandatory. Radio Access Network Authentication is by use of the A12 interface. Presently it is only applied in the Japan market. For the two methods, please refer to the figure 2 below.

There is a requirement to track the identity of the AT for regulatory purposes and since the outer CHAP cannot authenticate the AT, it is not enough. It turns out that in Japan, the ISP (internet service provider) and WSP (wireless service provider) may or may not be the same company-when the ISP is different from WSP.

- **Outer authentication between PDSN and AT**

The authentication in this layer is called Outer CHAP, which is transparent to AN, and performed between AT and PDSN. The relevant entities are :AT, PDSN, PDSN-AAA (RADIUS server). HA is also related for mobile IP.

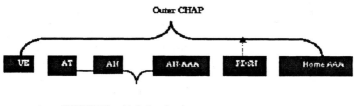

Figure 2.

The protocol between PDSN and AAA server is based on RADIUS. User authentication is based on network access identifier (NAI)[3] and password. Simple IP authentication is relevant to the entities MS, PDSN and PDSN-AAA server. PDSN acts as RADIUS[4] agent and PDSN-AAA as the server with the CHAP or PAP as the authentication protocol. Authentication flows for mobile IP is illustrated as below.

Mobile IP authentication is relevant to entities as AT, PDSN, AAA server and HA. Here the HA is responsible for the registration of the user address,

correlation of the user home IP address and COA (Care-of –Address)[5]. In the
same time, HA receives the data packet from MN and forwarded by FA to MN.
Authentication method is NN-FA Challenge- Response [RFC3012]. Figure 3

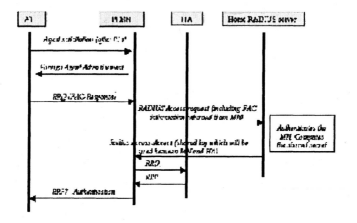

Figure 3.

shows the MIP authentication in registration. PDSN acts as FA and performs
authentication by use of PDSN-AAA. The main steps are :
1) After the establishment of PPP session, MN sends "Agent-Solicitation" message;
2) PDSN sends Agent-Advertisement to MN with the attachment of FA challenge;
3) If MN determines that it is in outside network then starts the MIP registration
and authentication;
4) AT calculates the response and resends RRQ to PDSN;
5) PDSN forwards to PDSN-AAA Access-Request message and attaches the
FAC and response.
6) PDSN-AAA forwards to HAAA and performs the authentication;
7) If authentication is successful then HAAA returns successful authentication
message with attachment of user HA address;
8) Once PDSN receives this, then forwards the user RRQ request to HA;
9) HA performs the AT authentication by use of the secret key issued by HAAA
and returns to PDSN RRP;
10) Once RRP is received, AT receives the HA address and performs re-
authentication by use of this address.
11) In the process of re-registration, PDSN may send directly the authentication
information to HAAA.

- **RAN layer authentication**

RAN layer authentication is also called inner CHAP, located between AT and AN through AN-AAA, which is optional. This authentication is initiated by AN with the objective to meet the user identity track by their operators and to support seamless handoff of the same packet data session between HRPD (1xEV-DO) and 3G1X with the IMSI return after this layer authentication, the IMSI is used in handoff. In this process, CHAP implements this layer authentication.

4.   Session security in the air interface

The key used by encryption and integrity agreement is generated by D-H protocol between AT and AN. There is no message encryption in version 1 until AES is proposed to use in the message and signaling encryption in version 2. But these versions do not apply the mandatory integrity protection to all the signaling. SHA-1 is optionally used to provide this protection.

5.   Security analysis and suggestion

5.1   Weak

Although current EV-DO security is basically perfect, it still exists some potential vulnerability from cryptography viewpoint. We shall analyze the security one by one in the following:

- Key generation is not based on authentication, but performs independently; hence the key can be easily thieved.

- The D-H is very easily exposed to man-in-the middle attack and very slow, high computation complexity.

- Session is not secure at all because integrity is not applied to any reverse access link and forward access link.

- No encryption is applied to any signaling in the air interface. So service is easily hijacked in the forward and reverse links.

- DoS attack is still unresolved. Even if the newest EV-DO security specification cannot protect against this attack.

5.2   Improvement

So this paper proposes the following suggestions based on the analysis above

- Key negotiation is not agreed by D-H algorithm. A secret ROOT key is pre-provisioned in the databases of the user equipment and core network. Then authentication and key agreement are performed based on this root key.

- Adopts the 3GPP UMTS AKA, which has been already proved to be very secure, this way, prevents man-in-the-middle attack and false key attacks based on the authentication. Two session keys are generated finally: cipher key CK and integrity key IK.

- Use the IPSec[6] or VPN to securely transport the key to other network entity after it is generated in the core network authentication center. Only the AT and authentication center can generate session key because only they know the ROOT keys.

- Enhanced SHA-1[7] in cryptography is used to integrity protection for signaling and wireless message with the application of the session key IK in the last step.

- Encryption is applied to both signaling and data in the access channel and transport channels by AES in the air interface [8].

6.   Conclusion

Compared to CDMA1X, authentication method and algorithm both adopt different ones. The up-to-date standard adopts the free AES without any export limit. The new specification can meet the security requirements of the user in authentication, encryption and key agreement process, then made up some important security vulnerabilities. But some famous vendor such as lucent has been already in many trials on it, which is very easy to implement and effectively protect the maximum interest of the current operator. This service protects the enough security as well as provides the high data rate service to users.

Acknowledgments

Special thanks to my dear advisor-Professor Xiao, who gave me very good enlightenment on cryptography with his deep knowledge. Also thanks to Bell-Labs which is a very good environment to pursue a career and many advanced technology to study.

References

[1] C.S0024 cdma2000 High Rate Packet Data Air Interface Specification V3.0, 2001

[2] RFC 1994: The challenge-Handshake Authentication Protocol (CHAP)

[3] RFC 2794, Calhoun, Perkins, Mobile NAI Extension March 2000.

[4] RFC 2138, Rigney, et al., Remote Authentication Dial In User Service (RADIUS), 1997.

[5] RFC 2002, Perkins, IPv4 Mobility, May 1995.

[6] FC 2401, Kent, Atkinson, Security Architecture for the Internet Protocol, November 1998.

[7] FIPS Publication FIPS 180-1, " Secure Hash Standard". April, 17, 1995.

[8] http://csrc.nist.gov/encryption/rijndael/rijndael-dos-refc.zip

A REMEDY OF ZHU-LEE-DENG'S PUBLIC KEY CRYPTOSYSTEM

Huafei Zhu

Institute for Infocomm Research
21 Heng Mui Keng Terrace
Singapore 119613.
huafei@i2r.a-star.edu.sg

Yongjian Liao

Department of Information and Electronics Engineering
Zhejiang University
Hangzhou, PR. China

Abstract   In 1998 Cramer and Shoup published a remarkable public key encryption scheme and we mentioned that the scheme involves a large sizes of secret keys. This leaves an interesting problem whether one can reduce the key sizes of the original Cramer and Shoup's scheme [4]. We made the first step to present a variation by defining the keys of Cramer-Shoup's test function as $c = g^x$ and $d = g^y$ in 1999 [9]. Unfortunately, the variation scheme was subsequently broken by Borst, Preneel and Vandewalle in 2000 [2]. Lucky enough, we are able to provide a remedy of Zhu-Lee-Deng's scheme in this short paper.

Keywords: decisional Diffie-Hellman assumption, standard complexity model, Zhu-Lee-Deng's scheme

1.   Introduction

Soon after Cramer and Shoup published a remarkable public key encryption scheme in 1998 [4], we mentioned that this public key cryptosystem involves a large sizes of secret keys. It is an interesting problem whether one can reduce the key sizes of the original Cramer and Shoup's scheme therefore. We made the first step to reduce the key sizes of the original Cramer and Shoup's scheme and presented a modification scheme by defining $c = g^x$ and $d = g^y$ in 1999 [9]. Unfortunately, the scheme was subsequently broken by Borst, Preneel and

Vandewalle in 2000 [2]. Fortunately, we are able to present a remedy scheme in this short paper.

To study the security of a newly developed cryptosystem, one could employ two standard models. 1) Random oracle model: In the random oracle paradigm setting, an ideally random and imaginary oracle, is assumed when one proving the security of cryptographic algorithms [3]. A random oracle H generates an answer randomly to the query posted to H at first, if the same query is asked later, H will answer the same value as was provided to the first query. The main advantage using of random oracle paradigm is that it can much more easily provide concrete security analysis, which avoids complexity theory and asymptotic theory. In practice, a random oracle is replaced by a random-like hash function such as SHA. We remark that all known cryptographic algorithms provably secure in the random oracle paradigm are very efficient and hence meeting for the practical requirements. However one must be caution that the schemes provably secure in the random oracle model do not imply that the schemes are also secure in the real world; And 2) Standard complexity model: In this circumstance, the related cryptographic primitives are based on standard assumptions, such as factoring problem and discrete logarithm problem together with its variations, e.g., computational Diffie-Hellman assumption, decisional Diffie-Hellman assumption. Definitely this kind of security is encouraged both from the point views of the theoretical research and the practice.

2.   Notions and Definitions

The security of a public-key encryption scheme is definitely related to the ability of adversaries and underlying assumptions. To define the ability of adversaries, three basic models are considered:

-Semantic secure: a public key encryption scheme is said semantic secure, which is first mentioned by Goldwasser and Micali [6], if an adversary should not be able to obtain any partial information about a message given its cipher-text.

-Secure against chosen cipher-text attack: a public key encryption scheme is said secure against chosen cipher-text attack (or lunch time attack or midnight attack), developed by Naor and Yung [7], if an adversary, who has access to the decryption oracle before a target cipher-text is given, is not able to extract any information of message.

-Secure against adaptive chosen cipher-text attack (an equivalent notion called non-malleable security against adaptive chosen message attack [5]): a public key encryption scheme is called secure against adaptive chosen cipher-text [8], if an adversary, who has access the decryption oracle even after the target cipher-text is given and the adversary can query the decryption oracle any cipher-text but the target cipher-text, is unable to extract any information about the message.

Our goal is to provide a public-key encryption scheme that is provably secure against adaptive chosen cipher-text attack in the standard intractability paradigm. The following notions and facts will be used to prove the security of our scheme.

Computational indistinguishability: Two families of distributions δ_1 and δ_2 are said to be computationally indistinguishable if no probabilistic polynomial time Turing machine distinguisher can decide which distribution it is sampling from with a probability of success non-negligibly better than random guessing.

-Fact 1: If δ_1 and δ_2 are computationally indistinguishable and δ_2 and δ_3 are computationally indistinguishable, then δ_1 and δ_3 are computationally indistinguishable.

-Fact 2: If δ_1 and δ_2 are computationally indistinguishable, then $\delta_1 \times \delta$ and $\delta_2 \times \delta$ are computationally indistinguishable for any independent distribution δ, where $\delta_1 \times \cdots \times \delta_k$, the productive distribution, is defined to be a distribution on k-tuples where the ith component is sampled according to the distribution δ_i.

The underlying primitive of our scheme is the hardness assumption of the decisional Diffie-Hellman problem. We therefore review the famous quadruple decisional Diffie-Hellman Problem below.

-The distribution R^4 of random quadruple $(g_1, g_2, u_1, u_2) \in G^4$, where g_1, g_2, u_1 and u_2 are uniformly distributed in G, where G is a large cyclic group of prime order q.

-The distribution D^4 of quadruples $(g_1, g_2, u_1, u_2) \in G^4$, where g_1 and g_2 are uniformly distributed in G while $u_1 = g_1^r$ and $u_2 = g_2^r$ are computed form an r which is uniformly distributed in Z_q.

An algorithm that solves the quadruple Decisional Diffie-Hellman problem is a statistical test that can efficiently distinguish these two distributions. Decisional Diffie-Hellman assumption means that there is no such a polynomial statistical test. This assumption is believed to be true for many cyclic groups, such as the prime sub-group of the multiplicative group of finite fields. To prove the security of our scheme, we also make use of the following Lemma, which is proved below.

Lemma [1]: Two distributions defined below are indistinguishable under the sole assumption of the standard quadruple Decisional Diffie-Hellman problem:

-The distribution R^{2k} of any random tuple $(g_1, \cdots, g_k, u_1, \cdots, u_k) \in G^{2k}$, where $g_1, \cdots, g_k$, and $u_1, \cdots, u_k$ are uniformly distributed in G^{2k};

-The distribution D^{2k} of tuples $(g_1, \cdots, g_k, u_1, \cdots, u_k) \in G^{2k}$, where $g_1, \cdots, g_k$ are uniformly distributed in G^k while $u_1 = g_1^r, \cdots, u_k = g_k^r$ for an r uniformly distributed in Z_q.

3.   Our remedy scheme

The original Zhu-Lee-Deng's scheme is defined as follows (please refer to [9] for further reference): Let H be a family of collision free hash functions. Let G be a group of prime order q for which the discrete logarithm problem is intractable and let g be a generator of G. The private keys are given by a pair $x, y \in G$. The public key are $(c, d) = (g^x, g^y)$. To encrypt a message $m \in G$, the protocol goes as follows:

- Choosing $r \in z_q$ at random, computing $u = g^r$, $v = mc^r$, $\alpha = H(u, v)$ and $\beta = c^\alpha d^r$. The cipher-text is (u, v, β).

- Given a putative cipher-text (u, v, β), computing $\alpha = H(u, v)$, and testing whether $\beta = c^\alpha u^y$ holds. If it does not hold, the algorithm outputs $reject$; Otherwise the decryption algorithm outputs the message $m = v/u^x$.

Borst, Preneel and Vandewalle's attack is succeed as the intermediate value α defined in our test function can be isolated. The remedy cryptosystem is defined below.

-Key generation algorithm: Let p be a large safe prime ($p = 2q + 1$ and q is a large prime). Let $G \subset Z_p^*$ be a sub-group of order q. Let H be collision-free hash function with output range Z_q; On input 1^k, the key generation algorithm chooses $g_1 \in Z_p^* \backslash \{1\}$ with order q, and $w, x, y, z \in Z_q$ uniformly at random and computes $g_2 = g_1^w$, $c = g_1^x$, $d = g_1^y$ and $h = g_1^z$. The output of the key generation algorithm is a public and secret key pair. The private key is denoted by (w, x, y, z). The public key is denoted by (g_1, g_2, c, d, h, H).

-Encryption algorithm E: To encrypt a message $m \in Z_p$, the encryption algorithm chooses $r \in Z_q$ uniformly at random, then computes $u_1 = g_1^r$ mod p, $u_2 = g_2^r$ mod p, $e = mh^r$ mod p, $\alpha = H(u_1, u_2, e)$ and $v = c^r d^{r\alpha}$ mod p. The output (u_1, u_2, e, v) is defined the cipher-text of message m.

-Decryption algorithm D: Given a putative cipher (u_1, u_2, e, v), the decryption algorithm computes $\alpha = H(u_1, u_2, e)$, then tests whether the equations $u_2 = u_1^w$ mod p, and $u_1^{x+y\alpha} = v$ mod p hold. If the both equations are valid, then the decryption algorithm outputs $m = e/u_1^z$ mod p, Otherwise it outputs reject.

The definition of security First, the encryption scheme's key generation algorithm is run, with a security parameter as input. Next the adversary makes arbitrary queries to the decryption oracle D, decrypting the cipher-texts of it choice. Next the adversary chooses two message $m_0, m_1 \in Z_p$, and sends these to the encryption oracle E. The encryption oracle chooses a bit $b \in \{0, 1\}$, at random and encrypts m_b. The correspondent cipher-text is given to the adversary (the internal coin tosses of the encryption oracle, in particular b, are not in the adversary's view). After receiving the cipher-text from the decryption oracle, the adversary continues to query the decryption oracle, subject only to the restriction that the query must be different than the output of the encryption oracle. At the end of game, the adversary outputs $\acute{b} \in \{0, 1\}$, which is supposed to be the adversary's guess of the value b. If the probability that $\acute{b} = b$ is $1/2 + \epsilon$, the adversary's advantage $Adv(D) := \epsilon$; We say a public key cryptosystem is secure against adaptive chosen cipher-text attack, if the adversary's advantage $Adv(D) = \epsilon$ is an negligible amount.

Theorem: The public key cryptosystem defined above is secure against adaptive chosen cipher-text under the joint assumption of decisional Diffie-Hellman problem as well as the assumption of the existence of collision free hash function H.

Proof: Given a quadruple (g_1, g_2, u'_1, u'_2) which is either a random quadruple or the Diffie-Hellman quadruple, we want to construct a distinguisher D so that it is able to distinguish whether it comes form random quadruple or Diffie-Hellman quadruple with non-negligible advantage with the help of the adversary who is assumed to be able to break the public key cryptosystem described above with non-negligible probability. We allow the adversary chooses two message $m_0, m_1 \in Z_p$, and sends these to the encryption oracle described below. The encryption oracle chooses a bit $b \in \{0, 1\}$, at random and encrypts m_b. The correspondent cipher-text is given to the adversary. After receiving the cipher-text from the decryption oracle, the adversary continues to query the decryption oracle, subject only to the restriction that the query must be different than the output of the encryption oracle.

The construction of simulator on input (g_1, g_2, u'_1, u'_2) is described as follows:

-Key generation algorithm KG_1: Let G be a sub-group of prime order q. We chosen $x_1, x_2, y_1, y_2, z_1, z_2 \in Z_q$ at random and computes $c = g_1^{x_1} g_2^{x_2}$, $d = g_1^{y_1} g_2^{y_2}$ and $h = g_1^{z_1} g_2^{z_2}$. The private key is $(x_1, x_2, y_1, y_2, z_1, z_2)$ and the public key is (g_1, g_2, c, d, h, H), where H is a collision free hash function with output range Z_q.

-Encryption oracle E_1: Given (g_1, g_2, u'_1, u'_2), m_0 and m_1, it chooses a random bit $b \in \{0, 1\}$ uniformly at random, and then computes $\acute{e} = m_b u'^{z_1}_1 u'^{z_2}_2$, $\alpha' = H(u'_1, u'_2, e')$ and $v' = u'^{x_1+y_1\alpha'}_1 u'^{x_2+y_2\alpha'}_2$. The output of E_1 is the cipher-text (u'_1, u'_2, e', v') of message m_b.

-Decryption oracle D_1: Given a putative cipher-text (u'_1, u'_2, e', v'), it computes $\alpha' = H(u'_1, u'_2, e')$, and tests whether $u'^{x_1+y_1\alpha'}_1 u'^{x_2+y_2\alpha'}_2 = v'$, if this condition does not hold, the decryption algorithm outputs reject; otherwise, it outputs $m_b = e'/u'^{z_1}_1 u'^{z_2}_2$.

we consider the following two cases:

Case 1: If (g_1, g_2, u'_1, u'_2) is a random quadruple. We want to show that there is no information leaked . In fact, our simulator is the same thing as the simulator of Cramer-Shoup's public key encryption scheme [4]. With the same argument as the Lemma 2 in [4], we know that when the simulator's input is a random quadruple, the distribution of the hidden bit b essentially independent from the adversary's view.

Case 2: If (g_1, g_2, u_1', u_2') is the Diffie-Hellman quadruple, we want to show that the adversary's advantage on (KG_1, E_1, D_1) is the same as that in (KG_2, E_2, D_2), which is defined below:

-Key generation algorithm KG_2: Let p be a large safe prime ($p = 2q+1$ and q is a large prime). Let $G \subset Z_p^*$ be a sub-group of order q. Let H be collision-free hash function with output range Z_q; On input 1^k, the key generation algorithm chooses $g_1 \in Z_p^*\backslash\{1\}$ with order q, and $w, x, y, z \in Z_q$ uniformly at random and computes $g_2 = g_1^w$, $c = g_1^x$, $d = g_1^y$ and $h = g_1^z$. The output of the key generation algorithm is a public and secret key pair. The private key is denoted by (w, x, y, z). The public key is denoted by (g_1, g_2, c, d, h, H).

-Encryption algorithm E_2: To encrypt a message $m_b \in Z_p$, the encryption algorithm computes $e' = m_b u_1'^z \mod p$, $\alpha' = H(u_1', u_2', e')$ and $v' = u_1'^{x+\alpha'y} \mod p$. The output (u_1', u_2', e', v') is defined the cipher-text of message m_b.

-Decryption algorithm D_2: Given a putative cipher (u_1', u_2', e', v'), the decryption algorithm computes $\alpha' = H(u_1', u_2', e')$, then tests whether the equations $u_2' = u_1'^w \mod p$, and $u_1'^{x+y\alpha'} = v' \mod p$ hold. If the both equations are valid, then the decryption algorithm outputs $m_b = e'/u_1'^z \mod p$, Otherwise it outputs reject.

We show that the two games (KG_1, E_1, D_1) and (KG_2, E_2, D_2) are equivalent up to the point where any invalid cipher-text can be rejected except for a negligible amount according to the following argument.

Indeed, since the decryption algorithm in game (KG_2, E_2, D_2) knows the trapdoor information $w = \log_g h$, we can assume that (g_1, g_2, u_1, u_2) is always from the Diffie-Hellman quadruple. As the decryption algorithm in the game (KG_1, E_1, D_1) is able to reject any invalid cipher-text except for negligible amount (the same argument as Lemma 1 presented in [4], where a cipher-text (u_1, u_2, e, v) is called valid if $\log_{g_1} u_1 = \log_{g_2} u_2$), it follows that the two games are equivalent up to the point where an invalid cipher-text is not rejected (however, the probability that this happens is negligible).

Furthermore we show that the adversary's advantage in game (KG_1, E_1, D_1) and in game (KG_2, E_2, D_2) are same. Indeed, the adversary's attack is restricted to adaptive chosen valid cipher-text attack to the game (KG_1, E_1, D_1) and to the game (KG_2, E_2, D_2). The distribution of valid cipher-texts (u_1', u_2', e', v') in game (KG_1, E_1, D_1) is denoted by δ_1 while the distribution of valid cipher-texts (u_1', u_2', e', v') in game (KG_2, E_2, D_2) is denoted by δ_2. Since (g_1, g_2, u_1', u_2') is Diffie-Hellman quadruple, it follows that $(g_1, g_2, h, u_1', u_2', u_1'^z)$ is the Diffie-Hellman six-tuple generated by game (KG_2, E_2, D_2) while $(g_1, g_2, h, u_1', u_2', u_1'^{z_1+wz_2})$ is the Diffie-Hellman six-tuple generated by game (KG_1, E_1, D_1)

according to the lemma presented in section 2. Consequently the distribution δ_1 is statistically indistinguishable to the distribution δ_2. It follows that the adversary's advantage in game KG_1, E_1, D_1) differs from game (KG_2, E_2, D_2) at most by an negligible amount.

Now we can build a distinguisher which is able to distinguish a random quadruple and the Diffie-Hellman quadruple with non-negligible advantage as follows.

Given (g_1, g_2, u_1', u_2'), we run decryption oracle of the simulator (KG_1, E_1, D_1), and the adversary will output a bit b' eventually. If $b' = b$, the distinguisher outputs a bit 1 indicating (g_1, g_2, u_1', u_2') chosen from Diffie-Hellman quadruple, otherwise, it outputs 0. By assumption, the adversary is able to guess the correct value b with non-negligible advantage, This immediately implies a statistical test distinguishing random quadruple from Diffie-Hellman quadruple.

4.   Conclusions

We have presented a remedy scheme provably secure under the hardness assumption of decisional Diffie-Hellman problem as well as the existence of collision free hash function H in the standard complexity model. We should point out here this scheme although defined over Z_p, can be easily extend to be defined over a set of elliptic curves where the decisional Diffie-Hellman assumption is reserved.

Acknowledgments

The research supported by NSFC 60273058.

References

[1] F.Bao. Robert.H.Deng, H.Zhu. Variations of Diffie-Hellman problem. ICICS 2003, Springer-Verlag, 2003.

[2] J.Borst, B.Preneel, J.Vandewalle. Comments on Variation of Cramer-Shoup public key scheme. Electronics Letters. Page(s): 32. Volume: 36, Issue: 1, Year: 6 Jan 2000.

[3] M.Bellare and P.Rogaway: Random Oracles are Practical: A Paradigm for Designing Efficient Protocols. ACM Conference on Computer and Communications Security 1993: 62-73.

[4] R.Cramer and V.Shoup. A Practical Public Key Cryptosystem Provably Secure against Adaptive Chosen Ciphertext Attack. In Crypto '98, LNCS 1462, pages 13-25, Springer-Verlag, Berlin, 1998.

[5] D.Dolev, C.Dwork, M. Naor: Nonmalleable Cryptography. SIAM J. Comput. 30(2): 391-437 (2000)

[6] S.Goldwasser, S.Micali. Probabilistic encryption. Journal of computer system and science. Vol.28, 270-299, 1984.

[7] M.Naor, M.Yung. Public key cryptosystem secure against chosen cipher-text attacks.22nd Annual ACM Symposium on the theory of computing, 1990, 427-437.

[8] C.Rackoff,D.Simon. Non-interactive zero-knowledge proof of knowledge and chosen cipher-text attacks. Cryptology-Crypto'91. 433-444, Springer-Verlag, 1992.

[9] H.Zhu; Lee Chan; X. Deng. Variation of Cramer-Shoup public key scheme, Page(s): 1150, Electronics Letters Volume: 35, Issue: 14, Year: 8 Jul 1999.

QUANTUM CRYPTOGRAPHIC ALGORITHM FOR CLASSICAL BINARY INFORMATION

Nanrun Zhou, Guihua Zeng

Department of Electronic Engineering,
Shanghai Jiaotong University,Shanghai 200030, China
znr21@sjtu.edu.cn, ghzeng@sjtu.edu.cn

Abstract   Based on quantum computation, a novel quantum cryptographic algorithm that can be used to encrypt classical message is proposed. The security and the physical implementation of this algorithm are analyzed in detail. It is shown that the algorithm can prevent quantum attack strategy as well as classical attack strategy. There are multiple advantages in using the proposed algorithm, most important of which is that it can be implemented with the current technology.

Keywords: cryptology, quantum cryptographic algorithm, quantum computation

Introduction

Quantum cryptography is of particular interest since the initial proposal of quantum key distribution in 1984 (Bennett 1984) and its experimental demonstration in 1992 (Bennett 1992). Current investigations of quantum cryptography are mainly concentrated on there aspects: quantum key distribution (Bennett 1992 1989, Erkert 1991, Bennett 1992, Brandt 2003), quantum secret sharing (Hillery 1999, Tyc 2002, Tittel 2001) and quantum cryptographic algorithm (Zeng 2002, Boykin 2003). The goal of quantum cryptographic algorithm and classical cryptographic algorithm is consistent, i.e. to protect secret information or keep communications private. The difference between quantum and classic cryptographic algorithm is as follows: the former is based on quantum laws while classical cryptography is based on pure mathematic principles. With the vast progress made in quantum computation, the possible quantum computer poses a threat to the classical cryptosystem in principle (Nielsen 2000). For example, the powerful Shor's quantum factoring algorithms for factoring and discrete logarithm and quantum Grover's searching algorithm are subtly designed according to the principle of quantum mechanics. How to devise an algorithm to resist quantum attacks is an important issue in data protection.

Fortunately, the currently proposed quantum cryptology is thought to be helpful not only because the existence of the eavesdropper can be detected in the quantum situation but also because the non-orthogonal quantum states can't be reliably distinguished (Nielsen 2000). In this sense, to cope with the possible and powerful quantum computer, quantum cryptology is one of the best candidates. In addition, an unconditionally secure algorithm in practical is also significant to the classic information protection.

To our knowledge, there are still not many good and feasible quantum cryptographic algorithms proposed at the nowadays technology level. On the one hand, no classical algorithm to date is both theoretically secure and practical. On the other hand, even if the quantum computer comes true some day, it is not necessary and possible to transfer all the existing secret classical information in the form familiar to ordinary people into quantum information, nor does the pre-shared classical keys as long as the security can be guaranteed. From this point of view, in this paper we propose a novel and practical quantum cryptographic algorithm.

1.   Quantum cryptographic algorithm

In the quantum situation, most algorithms are based on communicators pre-sharing quantum states, such as the EPR pair, which is impossible with the existing technology since quantum memory remains an open technology challenge. Motivated by these, we present a novel quantum cryptographic algorithm. It requires communicators to pre-share four groups of keys. Encryption is implemented by quantum computation, which can be realized by current technology. Because each encryption process is under the control of the key and the final ciphertext states are non-orthogonal, the eavesdropper cannot acquire fixed ciphertext without the keys, so eavesdropping attack is invalid, which is guaranteed by the no-cloning theorem in quantum mechanics (Wooters 1982). Similarly, the non-orthogonality of the ciphertext renders the Trojan horse attack strategy impossible (Gisin 2002).

1.1   Encryption process

Let us consider the encryption of the ith classical plaintext bit using the corresponding i^{th} key element of each group of keys. If the keys are used up, reuse the pre-shared keys from the beginning. The detailed encryption process is as follows.

Step 1: Preparation. Alice prepares the quantum ancilla state $|k_1^1 k_1^2\rangle$ according to the first group of key element k_1, where k_1^1 and k_1^2 are two key elements of the ith key pair in k_1 (For simplicity, in the context the i is left out). When the classical key element pair are 00, 01, 10 and 11, Alice prepares the corresponding quantum states $|00\rangle$, $|01\rangle$, $10\rangle$ and $|11\rangle$, respectively. Provided the classical binary message bit to be encrypted is m, Alice prepares the message quantum state $|m\rangle$ and generates the tensor product state $|k_1^1 k_1^2 m\rangle$ of the ancilla quantum state and the message quantum state. The result C_1 of this step is

$$C_1 = \{|k_1^1 k_1^2 m\rangle | k_1^1, k_1^2, m \in \{0, 1\}\} \tag{1}$$

where $\{...\}$ denotes a set.

Step 2: Controlled-NOT operation. Alice performs a Controlled-NOT operation ac- cording to the second group of key element k_2. The classical key elements 0 and 1 represent taking the first qubit and the second qubit as the control qubit, respectively. The third qubit (message qubit) always acts as the target qubit. After processing Alice gets the result $C_{k_2+1,3}|k_1^1 k_1^2 m\rangle$, where $C_{k_2+1,3}$ represents the Controlled-NOT gate with the $k_2 + 1$ qubit as the control qubit and the third qubit as the target qubit, respectively. Therefore, the resulting ciphertext states can be formulated as

$$C_2 = \{|k_1^1 k_1^2 \alpha_m\rangle | k_1^1, k_1^2, k_2, m \in \{0, 1\}\} \tag{2}$$

where $\alpha_m = |\alpha\rangle_m = [(\delta_{0,k_2} k_1^1 + \delta_{1,k_2} k_1^2) \oplus m]_m$ and the subscript m denotes the bit related to the original message bit. The third qubit in each state of C_1 is the original information qubit, but in each state of C_2 it is the result of the Controlled-NOT transformation and is no longer the original information qubit itself.

Step 3: Permutation. The existing algorithms usually fix a qubit position to represent the private information qubit, which may pose threats to the security of the cryptographic system in some special cases. Actually, Alice can permute two qubits in the resulting state according to the third group of key element k_3. If k_3 is 0, leave the state alone, otherwise swap the second and the third qubits. The set of the possible ciphertext states C_3 is given as follows

$$C_3 = \{\{\delta_{0,k_3}|k_1^1 k_1^2 \alpha_m\rangle + \delta_{1,k_3}|k_1^1 \alpha_m k_1^2\rangle\} | k_1^1, k_1^2, k_2, k_3, m \in \{0, 1\}\} \tag{3}$$

Ciphertext states in C_3 are different in form from those in C_1, the second and the third qubits of each state of C_3 may involve information about the message (plaintext), unlike those of C_1 where the information about the plaintext is just confined to the third qubit. Thus the ciphertext space is doubled.

Step 4: Non-orthogonality. Up to now, the intermediate ciphertext states Alice obtained are orthogonal, which can be distinguished and aren't suitable for propagating on the channel. To overcome this weakness, Alice carries out quantum computation on the ciphertext states in C_3 under the control of the fourth group of key element k_4 in order to make the final ciphertext states non-orthogonal. If the key element is 00 or 11, then leave the third qubit of state in C_3 alone, i.e. the third qubit remains in the state $|0\rangle$ or $|1\rangle$; but if the key element is 01 or 10, certain computation needs to be made on the third qubit.

When the key element is 01, Alice applies H gate to the third qubit and the resulting output state will be $|+\rangle$ if the input state is $|0\rangle$, or $|-\rangle$ if the input state is $|1\rangle$. On the other hand, when the key element is 10, Alice applies ZH gate to the third qubit. The possible ciphertext states in C_4 obtained from this step can be expressed as

$$
\begin{aligned}
C_4 &= \{(\delta_{00,k_4} + \delta_{11,k_4} + \delta_{01,k_4}H + \delta_{10,k_4}ZH)\{\delta_{0,k_3}|k_1^1 k_1^2 \alpha_m\rangle \\
&+ \delta_{1,k_3}|k_1^1 \alpha_m k_1^2\rangle\}|k_1^1, k_1^2, k_2, k_3, m \in \{0,1\}, k_4 \in \{0,1\}^2\} \quad (4)
\end{aligned}
$$

Thus, the whole encryption process comes to the end. Eq.(4) shows that the cipher-text states are non-orthogonal, and the message bits do not hide in the fixed position. Furthermore, the ciphertext space is doubled again.

1.2   Decryption process

The sequence of decryption process is right inverse of that of the encryption process. Because the above quantum operations are unitary, the decryption process can be completed easily under the guidance of the pre-shared keys.

2.   Security analysis

Firstly, let $|\psi_{ti}\rangle$ be the linear combinations of all the possible quantum states with equal probability in the ciphertext set C_t ($t = 1, 2, 3, 4$), which corresponds to the ith bit. The density matrices can be easily calculated and the results are $|\psi_{ti}\rangle\langle\psi_{ti}| = I$. The density matrix of n ciphertext states $|\psi_4\rangle$ related to the n bits classical message is

$$
|\psi_4\rangle\langle\psi_4| = \prod_{i=1}^{n} |\psi_{4i}\rangle\langle\psi_{4i}| = I \quad (5)
$$

Eq.(5) demonstrates that the ciphertext is homogeneous and includes no plaintext information. Therefore, the proposed quantum cryptographic algorithm is perfect privacy. Secondly, different ciphertext states are undistinguishable. One can calculate

$$
\langle\psi_{4i}|\psi_{4i}\rangle = (1 + \sqrt{2})/16 \quad (6)
$$

Eq.(6) explains that different ciphertext states are non-orthogonal. Due to the principle of quantum mechanics, the non-orthogonal states cannot be reliably distinguished. Our algorithm makes the ciphertext states non-orthogonal so that the ciphertext states are undistinguishable, which can prevent eavesdropping attacks. Thirdly, Trojan horse attack strategy gets no useful information (Gisin 2002).

3. Physical realization

The pre-shared four-group of keys can be realized by classical or quantum means. To guarantee the absolute security of the keys, one can employ quantum key distribution, e.g. BB84 protocol, which is now mature and commercially available (id Quantique). Encrypting procedure of the proposed algorithm is shown in Fig.1. In the figure, S_1, S_2 and S_3 are three quantum switches under the control of k_2, k_3 and k_4, respectively. When the element of k_2 is 0, S_1 switches onto 1, otherwise S_1 switches onto 2. When the element of k_3 is 0, S_2 puts through 1, or else 2. When the element of k_4 is 00 or 11, S_3 links to 2 and when the element is 01, S_3 links to 1, or else 3.

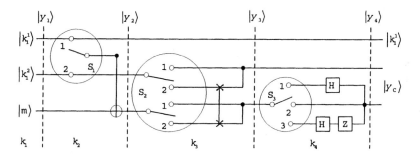

Figure 1. Schematic for encryption by quantum cryptographic algorithm

From Fig.1, one can see that the quantum cryptographic algorithm based on quantum computation only involves simple quantum logic gates, which can be realized physically easily. Furthermore, the message to be encrypted and the keys adopted while encrypting are all classical. The ciphertext states are directly sent to Bob through optical fiber channel or via air after all the encryption processes are finished by Alice. Upon receiving the ciphertext state, Bob decrypts it in time and it is not necessary to store the quantum states. Most important of all, the algorithm can be realized with the existing technology and may have wide application with the quantum error-correcting technology to enhance its performance.

4. Summary

In this paper, a novel quantum cryptographic algorithm to encrypt the classical binary bits is proposed. The security and the physical implementation of the quantum cryptographic algorithm are analyzed in detail. It is shown that the proposed algorithm can prevent quantum attack strategy as well as classical attack strategy. The circuit of encryption in principle is suggested and can be realized by the existing technology. Since there exists uncountable private classical information and it is impossible to transfer it into the form of quantum information even if the quantum computer comes true, the proposed algorithm is of much value.

Acknowledgments

This research is supported by National Nature Science Foundation of China (No. 60102001, 90104005).

References

[1] Bennett C.H., and Brassard G. (1984). *An update on quantum cryptography*, Advances in Cryptology: Proceedings of Crypto'84, pp.475. (Springer-Verlag)

[2] Bennett C.H., Bessette F., Brassard G., Salvail L., and Smolin J., (1992). *Experimental quantum cryptography.* J. Crypto., vol.5, pp.3.

[3] Bennett C.H., and Brassard G. (1989). *The Dawn of a new era for quantum cryptography: The experimental prototype is working*, Sigact news, vol.20, pp.78.

[4] Erkert A.K., (1991). *Quantum cryptography based on Bell's theorem*, Phy. Rev. Lett., vol.67, pp.661.

[5] Bennett C.H., (1992). *Quantum cryptography using any two nonorthogonal states*, Phys. Rev. Lett., vol.68, pp.3121.

[6] Brandt H.E. (2003). *Optimization problem in quantum cryptography*, J. Opt. B: Quantum Semiclass. Opt., vol.5, pp.S557.

[7] Hillery M., Buzek V., and Berthiaume A. (1999). *Quantum secret sharing*, Phys. Rev. A, vol.59, pp.1829.

[8] Tyc T., and Sanders B.C. (2002). *How to share a continuous-variable quantum secret by optical interferometry*, Phys. Rev. A, vol.65, pp.042310.

[9] Tittel W., Zbinden H. and Gisin N. (2001). *Experimental demonstration of quantum secret sharing*, Phys. Rev. A, vol.63, pp.042301.

[10] Zeng G., and Keitel C.H. (2002). *Arbitrated quantum-signature scheme*, Phys. Rev. A, vol.65, pp.042312.

[11] Boykin P.O., and Roychowdhury V. (2003). *Optimal encryption of quantum bits*, Phys. Rev. A, vol.67, pp.042317.

[12] Nielsen M.A., and Chuang I.L. (2000). *Quantum Computation and Quantum Information*,(Cambridge University press).

[13] Wooters W.K., and Zurek W.H. (1982). *A single quantum cannot be cloned*, Nature, vol.299, pp.802.

[14] Gisin N., Ribordy G., Tittel W., and Zbinden H. (2002). *Quantum cryptography*, Rev. Mod. Phys., vol.74, pp.145.

[15] id Quantique S A, http://www.idquantique.com

PRACTICAL QUANTUM KEY DISTRIBUTION NETWORK BASED ON STRATOSPHERE PLATFORM[*]

Jie Zhu,Guihua Zeng

Department of Electronic Engineering, Shanghai Jiaotong University, Shanghai 200030, China
shhzhujie@sjtu.edu.cn,ghzeng@sjtu.edu.cn

Abstract   A practical quantum key distribution network system based on stratospheric platform is proposed. And the feasibility of stratosphere quantum key distribution network is analyzed. As the length of the quantum communication channel in atmosphere has reached 23.4km, the proposed quantum key distribution network system has good prospects of practical applications.

Keywords: quantum key distribution; quantum cryptography; stratospheric platform; quantum relay

Introduction

Since the first quantum key distribution (QKD) protocol was proposed in 1984(BB84,Bennett et al 1984), quantum cryptography has received extensive research(Ekert 1991,Bennett et al 1992,Bennett et al 1993) as it provides unconditional security for the obtained key and allows successful detection of eavesdropping. Up to now, experiments show that transmission distance of quantum bits may reach 100km(optic.org 2003)in optical fiber and 23.4km(Kurtsiefer et al 2002) in atmosphere, which means that the practical QKD system become possible. Excited news is that some QKD products have been manufactured in Switzerland(News feature 2002). To practice the QKD in atmosphere, Aspelmeyer and his colleagues proposed a model to distribute quantum key based on satellites network by employing entangled photons(Aspelmeyer et al 2003). However, this way will not only burden high cost, but also be impossible according to the current technology.

A more practical model of distributing quantum key in atmosphere is to employ the stratospheric platform according to the current technology. In the

[*] This research is supported by National Nature Science Foundation(No. 60102001)

proposed model we employ stratospheric platform as either a transmitter, or a receiver, or a relay station in stratosphere quantum communication (SQC), the situation will be different. Normally, a stratospheric platform is floated in a 15km atmosphere band above the ground. The cost to build up a stratospheric platform is far below it to launch a satellite. At the same time, it is convenient to maintain the equipments, and the communication environment is comparatively simple and stable, besides that it will never bring garbage into space. Recently, the experiment result of quantum communication distance of 23.4km means that SQC is quite feasible.

In this paper, a practical model of QKD based on stratospheric platform is firstly proposed. The organization of this paper is as follows. First, some characters of stratosphere are introduced. Then, the architecture of QKD network based on stratospheric platform is proposed. Finally, the quantum communication network based on optical fiber and classical communication network are associated with the QKD. It is shown that the scheme proposed in this paper has good prospects in practical applications.

1.   Feasibility of stratosphere QKD network

The stratosphere normally located in 10-15km above the earth surface, lying between the troposphere and mesosphere (near 45-50km altitude). Little moisture enters the stratosphere, so clouds are rare and violent storms don't occur there. Thus the condition is favor of the transmission of entangled photons in atmosphere. The stratospheric platform we mention here usually prefers to the airship which is lighter than air. People can perform military scout, science researching ,wireless communication and weather survey by using such airship(Shields 2003).

According to the characteristics of stratosphere and quantum communication, we recently proposed a novel model called as SQC (Stratosphere Quantum Communication). The SQC exploits quantum communication equipment and other service load embarked on the stratosphere airship platform as communication relay station or communication terminal. By exploiting the SQC communicators may communicate information such as obtaining quantum key among SQC stations and ground quantum communication (GQC) station via quantum channel or hybrid channel combining quantum channel with classical channel. Since the stable atmospheric properties of stratosphere and the stratospheric platform is only about 15km above the earth's surface. In addition, the quantum communication within 23.4km distance in experiment has been realized. Stratosphere QKD network is feasible.

2.   Models of QKD network

To realize the QKD network based on stratosphere platform, we first give some typical models of QKD network, which mainly include end-to-end QKD system, star-network of QKD, arbitrary QKD network and global QKD network. During our explanation of stratosphere QKD network, a famous QKD protocol, called as EPR protocol(Ekert 1991), is used as an example. And polarization-entangled photons will be employed here as the source of communication, which can be denoted as follows:

$$|\Psi\rangle = \frac{1}{\sqrt{2}} (|\uparrow\rangle_a |\downarrow\rangle_b - |\downarrow\rangle_a |\uparrow\rangle_b) \qquad (1)$$

where subscripts a, b denote Alice and Bob's particles in the same EPR pair.$|\uparrow\rangle$ and $|\downarrow\rangle$ are the possible states they maybe.

2.1   End-to-end QKD system

An end-to-end QKD system is shown in Fig.1. Either a SQC station or a GQC station can be a transmitter and the other will be a receiver.

A transmitter, say, Alice, includes a photon source for emitting pairs of spin$^1/2$ particles in a singlet state. The transmitter holds one photon of the entangled pair and make the other fly apart towards the receiver, Bob. Alice and Bob select the orientation of detectors randomly and independently for each pair of incoming particles. The results of their measurement are either +1 or -1, which means spin up or spin down, and this can imply one bit of information. After the transmission has taken place, Alice and Bob announce

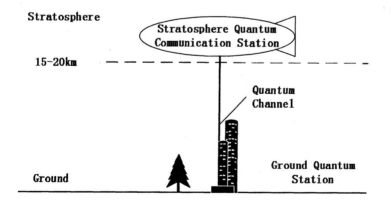

Figure 1.   End-to-end QKD system

publicly the orientations of the analyzers they have chosen, as stated by Ekert. The measurements are divided into two separate groups by Alice and Bob: a first group for which they used different orientations of their detectors, and a second group for which they used the same. They discard all the measurements in which either or both of them failed to record a particle at all.

Subsequently, Alice and Bob can reveal in public the results they got but only within the first group of measurements, which makes it determined whether the results of the second group owned by Alice and Bob are anticorrelated and can be converted into a secret string of bits or not. This secret string of bits is the so-called secret key, which can be used to perform secure communication.

As the process is fulfilled by physical means and is protected by the completeness of quantum mechanics, the eavesdropper cannot get any information from the particles during the transmission and his disturbance will be successfully detected.

Also, Alice and Bob can transfer entangled photons to each other so as to build up one or more quantum channels to perform QKD protocols. Such a QKD system is the most simple model in stratosphere QKD network.

2.2   Star-network of QKD

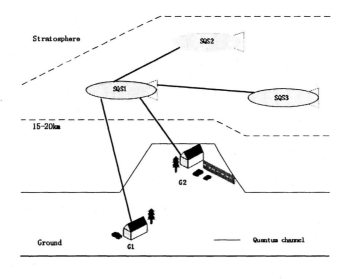

Figure 2.   Star-network of QKD

A little more complicated network is shown in Fig.2, which is called star-network of stratosphere QKD. Any quantum communication station can be a center node, so it can distribute entangled photons to any other nodes, including SQC stations and GQC stations, to set up a star-network of SQC.

Such a network can easily perform entanglement-based QKD protocols. Center will be a resource which is responsible for producing entangled photon pairs. Center distribute entangled particles to the legitimate users respectively, say, Alice and Bob. Then Alice and Bob choose the orientation randomly while detecting the incoming photons. They announce their results of measurement by classical stratospheric communication means, and deduce the secret key to complete encryption and decryption.

The star-network allows to perform QKD among multi-user. But it also has a disadvantage that once the center is unavailable, the network is hardly to proceed any quantum communication. With existing technologies, the above two QKD systems can be realized in practice.

2.3   Arbitrary QKD network

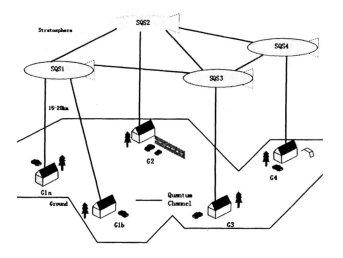

Figure 3.   Arbitrary QKD network

If relay technologies are adopted here, an arbitrary network of stratosphere QKD can be built up, as shown in Fig.3. The relay technologies mainly contain quantum teleportation, quantum purification and quantum swapping. With relay technologies, any node in network can be used to carry either a transmitter of entangled photons, or a receiver, or a relay station to distribute photons to further locations, which will permit different applications, but the relay module only redirects and/or manipulates qubit states without actually detecting them. The required shared entanglement can be established either by two downlinks or by using additional stratospheric platforms. For example, SQS1 sends one photon of an entangled photon pair to G1a and sends the other to SQS2. SQS2 sends one photon of an entangled pair produced by himself to G2, but keep down the other. Then SQS2 performs a Bell-state measurement on the two independent photons, of which one is kept down by SQS2 and the other is received from SQS1. The result of such measurement leads to that the two photons, one owned by G1a and the other owned by G2, become a new entangled photon pair. Subsequently, G1a and G2 may use the entangled photons to perform EPR protocol.

2.4   Global QKD network

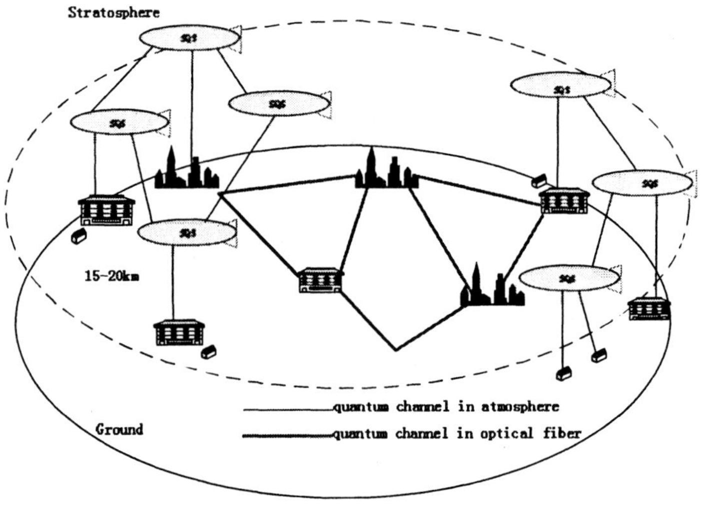

Figure 4.   Global QKD network

The most aggressive quantum QKD network is shown in Fig.4, in addition, the quantum network based on optical fiber and classical communication network are involved in such network. Some desirable attributes in the QKD network are safe management of keys, quantum authentication, efficient transmission of keys and robustness etc. Furthermore, such QKD can be integrated with IPsec so as to secure internet traffic, which has been stated in detail by Chip Elliott(Elliott 2002).

3. Implementation and applications

The realization of stratosphere QKD network depends on the technology of stratospheric platform. As stratospheric communication has many advantages compared to satellite communication, such as that its cost is relatively much lower, the maintenance is simple and convenient and it is in favor of environment protection. United States, Europe and Japan have invest much into the study and research of stratospheric platform. Once stratospheric platform can be practically performed(which is anticipated to be at the end of 2005) and is in wide use, our scheme is undoubtedly more feasible than the idea to distribute entangled photons using satellites. And the performance of stratosphere QKD network is also related to the weather condition. We are now studying some related problems and getting along. Moreover, relay technology should be required in order to form a wide-scale network, in which researchers have also made breakthroughs recently(Marcikic et al 2003,Grosshans et al 2003,Pan et al 2003)With the above conditions, stratosphere QKD, quantum authentication and quantum secret sharing can be performed. If stratosphere QKD network is combined with quantum communication network based on optical fibers and internet or other classical communication, a global secure communication network can be carried out.

4. Summary

Based on stratospheric platform, the architecture of a practical QKD network is presented here. It is shown that in good weather condition, the stratosphere QKD scheme is feasible and practical. And if we combine the stratosphere QKD network with optical quantum communication network and internet or other classical communication means, a global secure communication network can be implemented.

References

[1] C.H.Bennett and G.Brassard.*Quantum cryptography: public key distribution and coin tossing*.Proceedings of IEEE International Conference on Computers, Systems and Signal Processing. Bangalore India, 175£-179,1984.

[2] A.K.Ekert. *Quantum cryptography based on Bell's theorem*. Phys.Rev.Lett.67(6): 661-663,1991.

[3] C.H.Bennett, G.Brassard and N.D.Mermin. *Quantum cryptography without Bell's theorem*. Phys. Rev.Lett.68(5):557£-559,1992.

[4] C.H.Bennett, G.Brassard, C.Crepeau, R.Jozsa, A.Peres, and W.K.Wootters. *Teleporting an unknown quantum state via dual classical and Einstein-Podolsky-Rosen channels*. Phys. Rev. Lett. 70(13):1895-1899,1993.

[5] A.Pankine, K.Aaron, M.Heun, K.Nock, W.Wiscombe, B.Mahan, W.Su.*Geoscience and Remote Sensing Symposium*.IEEE International, 1:362£-364,2002.

[6] C.Kurtsiefer, P.Zarda, M.Halder, H.Weinfurter, P.Gorman, P. Tapster, and J.Rarity. *A step towards global key distribution.* Nature,419:450,2002.

[7] News Feature. *Can you keep a secret?* Nature,418:270£-272,2002.

[8] M.Aspelmeyer, T.Jenewein, A.Zeilinger, M.Pfennigbauer and W.Leeb. *Long-Distance Quantum Communication with Entangled Photons using Satellites.*quant-ph / 0305105,2003.

[9] A.Shields. http://optics.org/articles/news/9/6/3/1, Conference on Lasers and Electro-Optics. Baltimore US,2003.

[10] C.Elliot. *Building the quantum network.* New Journal of Physics,4:46.1£-46.12,2002.

[11] I.Marcikic, H.de Riedmatten, W.Tittel, H.Zbinden, and N.Gisin. *Long-distance teleportation of qubits at telecommunication wavelengths.* Nature,421: 509£-513,2003.

[12] F.Grosshans et al. *Quantum key distribution using Gaussian-modulated coherent states.* Nature,421:238£-241,2003.

[13] J.W.Pan, S.Gasparoni, M.Aspelmeyer, T.Jennewein, and A.Zeilinger. *Experimental realization of freely propagating teleported qubits.* Nature,421: 721£-725,2003.

A SURVEY OF P2P NETWORK SECURITY ISSUES BASED ON PROTOCOL STACK*

ZHANG Dehua

College of Elec. Eng., National University of Defense Technology, 410073, Changsha, China
zhangdh@nipc.org.cn

ZHANG Yuqing

National Computer Network Intrusion Protection Center, GSCAS, 100039, Beijing, China
zhangyq@nipc.org.cn

Abstract   People pay more attention to the security issues of the P2P Network, and propose different secure settlements to different applications. In this paper, we begin with studying the features of P2P Network, analyze the secure issues of the P2P Network, and establish the hierarchical division of P2P Network. We also analyze the secure issue according to the hierarchical division. In the end of this paper, we analyze the shortness of the existing security technology being used in the P2P Network applications, and give out the P2P Network security research orientation and tendency

Keywords: P2P network, network security

1.   Introduction

The P2P Network and its applications are the focus of the network application researches at this moment. Since Napster in 2000[2], people developed different applications soon afterwards. Like other network applications, the secure issue becomes the core issue that P2P Network has to be resolved. Although in SETI@home [5] the node assessing means was employed, some nodes have has been discovered to deceive in the system, and this also causes entire system security to be intimidated. People use different secure schemes according to the different secure demands. Most existing security settlements use the common security technology as reference, such as like PKI, SSL[11], yet the consummate

*This work is supported by National Natural Science Foundation of China under Grant 60373040 and National "863" Program Grant 2003AA142150.

secure settlement has not been proposed aimed at P2P Network. There is chiefly short of completely and systematic knowledge of the P2P Network security issue.

To the best of our knowledge, researches on security issues of P2P Network focus on the concrete application. And there is lack of the security definition and consideration of P2P Network as a whole. Here we analyze the security issues of P2P Network from the point of view of protocol stack. We also point out the security issues in different layers and the security settlements.

This paper introduces the basic concepts of the P2P Network and its security issues. The concept of security hierarchical layers of the P2P Network in Ch2 is proposed. And we the security settlements of existing system beginning with the analyzing of the security demand of different applications in Ch3 sum up are summed up. Ch4 analyze the security issues faced in different layer, the existing security hidden trouble and probable attack measure. In Ch5 the focus of research in the P2P Network and development tendency are pointed out.

2.   Basic Concepts

2.1   The P2P Network

2.1.1   Definition of the P2P Network.   A concept which network is the P2P Network should been bounded firstly.   There are some definitions such as [1]. These definitions represented in the P2P Network nature through distinct aspects. Here we may sum up some substances that the P2P embodies: resources and network node are situated in the network fringe, the node has both client and server capacity, the node has independence address means are independence to DNS, communicating with each other immediately.

We will sufficiently consider these distinguishing features of the P2P's Network at the analysis of the P2P's Network security issues.

2.1.2   The P2P Networking Protocol Stack .   The P2P Network is the virtual network establishing on the basis of available network. Comparing with the traditional network application we think that the P2P's Network is hierarchy. Here are the functions and definitions of each layer of the P2P Network.

1) Connection layer: the orientation of the resources and message of route are provided in this layer. There are three main kinds of realization methods of connection layer at the moment:

• Centralization: This type of network takes Napster as representative. All searching of node and resources is wholly starting from the server.

• Network of discrete structure: No central server is used to register the resources places, but it has a steady structure. Concrete representative consists of : Chord [14], and Tapestry [15]

• The network of discrete non-structuring: the topology of the nodes is configured, and the time when the node joins and leaves the network is not fixed. JXTA is such a type.

2) Service layer

This layer mainly consists of the operation interfaces from the consumer to the resources, which we called service here. This layer provided service of relevant functions to application layer. These services can be divided into the core service and the ordinary serve.

3) Application layer

It is manly consisted of concrete the P2P application. P2P application should provide reasonable interface that is used for users to manage and fix the network, and the answer process to event and operation.

2.2   The P2P Network Security

2.2.1   Definition of Security .   Security itself is very wide in range, and it embodies the idea of the reliability of system, degree of trust and fault-tolerance. Security is a layer concept and the layers is corresponding to the hierarchy of the P2P Network. Here we give the security definition in each P2P layer:

Definition 1: Security of the connection layer may be expressed:

$P(S) = F(n,f,h)$;

Here P (S) is security of the connection layer, and n is the method to define the name space, and F acts as the mapping algorithm, and h is the reliability hypothesis of the connection layer.

Definition 2: the security of service layer is the security of the security service.

That service security is different according to the different security service chosen by the application.

Definition 3: The security of application layer is chiefly meeting of secure demand to the system.

Whether meeting the secure demand of a system is the significant judge criterion of the security of application layer and even of entire system.

Researching the security of the P2P Network from these three layers is our main standpoint.

2.2.2   Research Methods of the P2P Network Security .   The P2P Network is a network system essentially .The research means can lean from what is used commonly to a network system. We will analyze the existing P2P applications using the two methods separately, finding out the resolution element and settlement scheme.

1) Begin with the security demand , and mapping the demands to each layers ,and complotting security functions in each layer, and guaranteeing the system security as whole.

2) Beginning with analyzing the secure hidden danger of each layer, and providing pertinence modification means that these two methods are the most common methods used in system's security design.

3.   Secure Demands Analysis of the P2P Network

3.1   P2P Computing

The P2P computing is the applications in which the CPU clock is shared. In the P2P computing the task is decomposed into subtask being finished in different nodes, by which the resource in the fringe can be used. The P2P computing is a kind of distributed computing on nature. Representative ness of P2P computing is SETI@home,Megi[13].

Here we may divide the P2P computing to such several implementations step: The mission formation, mission decomposition and mission distribution and result referring.

We sum up the security demand from the item mentioned before:

1) To the code protection .The correctness, completeness and non-deny should been guaranteed in the process of subtask publishing.

2) The two-way authentication be carried on between the nodes . When completing a task, we should trace the situation of the completing.

3) To node protection . This is required that the nodes participate in the P2P computing should not been suffered with the security hidden danger.

4) To the result protection . The subtask is completed and the result referred to is safeguarded

5) To the mutual information protection . When the mutual information exists, it should been guaranteed.

3.2   Cooperation Computing

Cooperation Computing is the most popular P2P application .The most famous Cooperation Computing in P2P Network is Groove and Avaki. Their emphasis on different places: the Groove is chiefly used in the business limit , consisted of the file sharing and corresponds forthwith and white-board communication and so on ; The Avaki chiefly provide supplies to the moving code
.

The Groove proposes relatively the integrated secure settlement scheme. In the Groove, the concept of security sharing room is inducted. The sharing files and programs are included in the security sharing room.

The Cooperation Computing deal with the resource sharing between users, so the access control issue should be settled mainly. Concretely included:

1) role based access control mechanism. Frequently a resource is determined whether or not to be accessed by user, according to the role of the user.

2) separating Consumer , node and resources . This is because the same user can access the resources from different node.

3) authentication mechanism based on user's attribute. In Cooperation Computing people are concerned about frequently is that the man with which attribute can participate in the work, and not what man he is.

4) people in distinct organization and areas may carry on the authentication reciprocally . By this user can get consistent authentication through different fields.

3.3   File Sharing

File sharing is the original designing intention of the P2P Network.There are two main types of this application. One is aiming at the exchanging file, the other is aiming at the making anonymous publish.Among these two types, the latter type has more security demands including:

1) Guaranteeing that the transforming file is integrated and secret
2) Guaranteeing that the file can not be forged
3) Guaranteeing that the file can not be tampered
4) System has the ability of auti-tracing and auti-auditing
5) Publishing is transparence to users

4.   The P2P Network Security Hidden Danger and Attack

4.1   Connection Layer

The main functions of the connection layer consist of resource searching, locating and routing. It is the base of the whole P2P Network. The routing arithmetic is composed of: naming space of resource, addressing space of nodes, mapping arithmetic from naming space to addressing space.

4.1.1   Distributing of the Space.   There are three kinds of distributing arithmetic which all have security issues. Here we mainly find out the hidden security danger and countermeasure.

1)Distributing the address of the resources and nodes completely randomly

Completely random distributing method is the common way used by many P2P researching. This may create some hidden danger. The misfeasor node may choose a certain number as its serial number. He can even choose the number of other certain resources, by which he may imitate someone.

2)A solution is to get the serial number form CA node using encryption.

This solution makes the certificate bound to a node. This method is a direct enlargement of PKI. Of cause CA may bring the extendibility of system, and may single node failure issue. All of issue should been solved when using this method.

3)Another method is to using the public key directly

The creation of the public key is stochastic, so the node can get the serial number at the same time, which do not need certificate from CA. this method solve the extendibility issue of using CA.

4.1.2   Mapping from Resource to Address Space.   There are two mainly mapping arithmetic. One is based on the hypothesis that the network architecture is discrete structure. This arithmetic includes CAN, DHT, Pastry and so on. This arithmetic can get high resource searching efficiency. The other is based on the hypothesis that the network is discrete unstructured. Both of these two methods use the idea of DHT.

Now the security of the DHT is researched by [16]. The main opinions of these researches are:

1)Security of the DHT

The research of this field is based on the archives of Morris. He proposed the framework of enhancing the security of the P2P Network. He viewed the job of creating, maintaining the routing table and node searching as security task. He gave the potential attack to the P2P Network. But he only gave the qualitative conclusion. Wallach[18] provided a deeper conclusion. And he made the improvement of DHT. Using this method he can guarantee that if there are less 30% hostility node in the network, the 99% routing message can arrive correctly.

2)Researching on the type of attack

People find out many attack against the routing arithmetic such as DOS , Byzantium attack and Sybil attack. [22] analyze the reason why this attack raises and the different attacking form in every conditions. And [22] draw the conclusion that it is very difficult to guarantee the security of the ID.

3)Analyzing security of the application

Bellovin[21] analyze the security issue faced by Napster and Gnutella[10]. This security issue is caused by the PUSH operation of Gnutella. This operation can step through the firewall. Someone can create DOS attack using is operation.

4.2   Service Layer

The security of service layer is mainly pointed to the security of the security service provided in the P2P Network. The mostly used security service is authentication based on certificate. Another important secure issue is how to establish the security point. Security point is the interface referring the security issue. In the process of implement a P2P Network, establishing the security point is an important issue.

4.2.1   Security Service Protocol.   The security service protocol is based on the certificate, so the core issue in P2P Network security is the management of certificate.

The security issue of SSL protocol has been study deeply. Now we consider the security issue of SSL in the P2P Network.

The existing methods of getting and validating certificate include:

1)Based on CA: In the P2P Network, this method has the expandability issue and the cost using this method is very high. This situation does not fit the design of the P2P Network.

2)Getting certificate from itself: That is every node can give certificate to itself. This method mainly refers to the issue of validating and managing the certificates.

At present using and managing certificates are difficult issue to be solved in P2P Network.

4.2.2   Establishing the Security Point.   Now the main method to define the security point includes:

1)Single security point: This system always has little special security demands. Security point is established where users log in. And other systems have default authentication.

2) Multi-security point: This system establishes multi-security point according to the system's demands, such as megi, which establishes two security points in communication layer and user authentication.

How to establish the security point is different according to each application. One should firstly guarantee that the security point can not been run around. And many hidden danger is caused by this mistake.

4.3   Application Layer

The security of application layer chiefly is meeting of secure demand to the system. Here we consider the security of application layer as the code security

.

At present there is less research on code security in P2P Network. This is because it is difficult to validate the security of code itself, and many project solve security issue in under layer, the up layer only complete mutual operation with users referring to little security issue.

5.   Conclusion

We can view from the feature of P2P Network that its security issue is similar to other network applications. Some common security settlements have good effect in the P2P Network.

Leaning from the current research, we can draw the conclusion that transform between fields and standardization are the most important issue to be solved. Transform between fields in P2P Network refers to in different fields the authentication and access control method can be understood each other. An effective method to solve the issue is standardizing the protocol. At present we can lean form the trust management system and Globus standard protocol. We think this work may make deep influence in P2P Network.

References

[1] Dejan S.Milojicic, et al., HP Laboratories Palo Alto. Peer-to-Peer Computing.HPL-2002-57, March 8th, 2002

[2] Napster, http://www.napster.com/

[3] I.Clarke, et al., Adistributed anonymous information storage and retrieval system.

[4] Lance Olson, .NET P2P: Writing Peer-to-Peer Networked Apps with the Microsoft .NET Framework. Microsoft MSDN magazine

[5] SETI@home, http://setiathome.ssl.berkeley.edu

[6] Antony Rowstron, et al., Scalable, decentralized object location and routing for large-scale peerto-peer systems

[7] Bernard Traversat, et al., Sun Microsystems, Inc. Project JXTA Virtual Network. October 28, 2002

[8] Avaki, http://www.avaki.com

[9] Jerome Verbeke, et al., Framework for Peer-to-Peer Distributed Computing in a Heterogenous, Decentralized Environment.Sun Microsystems, Inc.

[10] Gnutella http://www.gnutella.com

[11] SSL http:// www.openssl.org/

[12] Groove .http://www.groove.com

[13] Peer-to-Peer Architectures and the Magi Open-Source Infrastructure http://www.endtech.com/pdfs/ETI%20P2P%20white%20paper.pdf, January 2002

[14] Stoica, I., et al., Chord: A scalable peer to peer lookup service for Internet applications. SIGCOMM'01, ACM, San Diego,California (2001)

[15] Zhao, B.Y., et al., Tapestry: An infrastructure for fault-resilient wide-area location and routing. Technical Report UCB//CSD-01-1141, U. C. Berkeley (2001)

[16] Sit, E., et al., Security considerations for peer-to-peer distributed hash tables. IPTPS '02, Cambridge,Massachusetts (2002)

[17] Castro, M., et al., Practical byzantine fault tolerance. OSDI'99, New Orleans,Louisiana (1999)

[18] Wallach, D.S., et al., Extensible security architectures forJava. Proc. of the 6th ACM Symposium on Operating System Principles,Saint-Malo, France (1997) 116–128

[19] Waldman, M., et al., A robust, tamper-evident, censorshipresistant, web publishing system. Proc. 9th USENIX Security Symposium, Denver, Colorado (2000) 59–72

[20] Hazel, S., et al., Achord: A variant of the Chord lookup service for use in censorship resistant peer-to-peer. Proc. for the 1st International Workshop on Peer-to-Peer Systems, Cambridge, Massachusetts (2002)

[21] Bellovin, S., Security aspects of Napster and Gnutella. In: 2001 Usenix Annual Technical Conference, Boston, Massachusetts (2001) Invited talk.

[22] John Douceur: The Sybil Attack In : www.cs.rice.edu/Conferences/IPTPS02/101.pdf

DDOS SCOUTER:
A SIMPLE IP TRACEBACK SCHEME

Chen Kai

Bell-labs Research China, Lucent Technologies, Beijing 100080, China

kaichen@lucent.com

Hu Xiaoxin

Dept of Computer Sci.& Eng., University of Electron. Sci. and Tech., Chengdu 610054, China

Hao Ruibing

Bell-labs Research China, Lucent Technologies, Beijing 100080, China

Abstract   Defense against distributed denial-of-service attacks is one of the hardest security problems on the Internet. Among those problems, the most difficult problem is to trace the attacks back to its origin for the attackers always use incorrect or spoofed IP addresses in the attack packets. In this paper, we propose a multi-edge marking scheme, which allow the victim to traceback to or near to the origin of the attackers with the help of the network administrator. The scheme features high performance efficiency and no false positive. Compared with the previous solutions, it has high precision and low computation overhead for victim to reconstruct the attack paths. Base on this marking scheme, DDoS Scouter is developed.

Keywords: DDoS attacks, IP traceback, packet marking

1.   Introduction

With the wide deployment of Internet, security problems become the extreme threat to the Internet society. Due to the stateless and destination IP address routing natures of Internet, the Denial of Service attacks (DoS) are the most reported one among the security problems. A denial-of-service attack (DoS) aims at denying a victim (host, router, or entire network) providing or receiving normal services in the Internet. Distributed denial-of-service attacks (DDoS), typically conducted by flooding network links with large amounts of traffic(which is the focus of the paper), consume the resources of a remote host or network, thereby denying or degrading service to legitimate users. Such attacks are among the hardest security problems to address because they are easy to implement, difficult to prevent, and very difficult to trace.

In general, there are two types of flooding attacks: direct attacks and reflector attacks. In a direct attack, an attacker arranges to send out a large number of attack packets directly toward a victim. Attack packet types can be TCP, ICMP, UDP, or mixture of them. Before launching a direct attack, an attacker first sets up a DDoS attack network, consisting of one or more attacking hosts, a number of masters or handlers, and a large number of agents. The attacking host is a compromised machine used by the actual attacker to scan for vulnerable hosts and to implant specific DDoS master and agent programs. With an attack network ready, the attacking host may launch a DDoS attack by issuing an attack command with the victim's address, attack duration, attack methods, and other instructions to the masters. Each master, upon receiving the instructions, then passes them to its agents for execution. A reflector attack is an indirect attack in that intermediary nodes(routers and various servers), better known as reflectors, are innocently used as attack launchers. An attacker sends packets that require responses to the reflectors with the packets' inscribed source addresses set to a victim's address. Without realizing that the packets are actually address-spoofed, the reflectors return response packets to the victim according to the types of the attack packets. As a result, the attack packets are essentially reflected toward the victim, and the reflected packets can flood the victim's link if the number of reflectors is large enough.

Because the results of DDoS attacks are serious financial disaster to the victim, many research results[3,4,5,10,11,12,13,14,16,17,20], which are aimed at preventing the DoS/DDoS attacks, have been obtained in the research field. They can be divided into three lines: attack prevention and preemption(before the attack), attack detection and filtering(during the attack) and attack source traceback and identification (during and after the attack).

Although, it is infeasible to use IP traceback to stop an ongoing DDoS attack, it could be very helpful in identifying the attacker and collecting evidence for post-attack law enforcement.

Among the above techniques, attack traceback and identification has been much considered recently. It can usually be carried out after or during a DDoS attack. IP traceback refers to the problem, as well as the solution, of identifying the actual source of any packet sent across the Internet without relying on the source information in the packet. Up to now, there are generally two type approaches to the IP traceback problem. One is for routers to record information about packets they have seen for later traceback requests[3,14], named logging. Another is for routers to send additional information about the packets they have seen to the packets' destinations via either the packets[5,16] or another channel, such as ICMP messages[10].

In Bellovin's proposed ITRACE scheme, routers, with a very low probability, send ICMP messages to the destinations of packets they have just forwarded[10]. For a high-volume flow, the victim will eventually receive ICMPs from all of the ITRACE routers along the path back to the attackers, revealing its location. Savage and colleagues proposed a different scheme, in which routers with considerably higher probability mark the packets they process with highly compressed information that the victim can decode in order to detect the edges traversed by the packets, again enabling recovery of the path back to the attacker[5]. However, the scheme runs into computational difficulties as the number of attackers increases. This problem is addressed by Song and Perrig by supplementing the scheme with the use of network topology maps[16]. Recently, Snoeren and colleagues developed a Source Path Isolation Engine(SPIE) that records sets of hashes of packets traversing a given router[3]. A victim can then locate the path of a given packet by querying routers within a domain for the set of hashes corresponding to the packet, providing that they issue the query soon enough after the packet was transmitted that the record of its presence is still available. SPIE has a major advantage in that it can facilitate traceback of even low volume flows.

There is a dilemma in designing the IP traceback scheme: time and space requirements. Among the above systems, the logging related schemes are said not practical because the storage requirement of the router is too high; the marking related schemes have the disadvantages of high time consumption in marking packets collection and attack paths construction.

In the victim's view, the quick response to the attack is much more desired. However, in order to reduce the marking space requirement, the well recognized schemes proposed in[5] deployed some code techniques which led to the inefficiency in packet collection and attack path reconstruction, especially for DDoS attacks. The direct result is that the victim has to endure longer attack.

To make the IP traceback technique more practice, it is necessary to make a tradeoff between the time and space requirements. In this paper, we proposed an on-demand probabilistic multi-edge IP marking technique to do IP traceback. It is designed that the marking enabled router only doing marking when it receives the marking instruction from the network administrator. In the scheme, the record route IP option is used to mark the router's(by which the packet is forwarded) IP addresses. In the record route IP option, several IP addresses can be recorded. So, an attack packet can carry a segment of the attack path with which it can improve process of the attack path reconstruction greatly.

Based on the proposed IP traceback technique, DDoS Scouter system is designed to prevent the DDoS attacks. The system consists of attack detection, IP traceback and packet filtering(intelligent packet filtering using the traceback scheme will be studied in the other paper).

The rest of the paper is organized as follows. Section 2 proposes the basic and authenticated multi-edge marking and traceback algorithms. In section 3, the DDoS Scouter system is presented. The testing results of the system is described in section 4. In section 5, some problems of the system are discussed and it is concluded in section 6.

2.   Multi-edge marking

Keeping in mind that the IP protocol should not be modified or just little modification should be made when an IP marking scheme was designed or developed. In addition, the designed or developed system should not add too much process burden to the routers, which would lower the performance of the routers. It is also noticed that, without the help of the network administrator, it is very inefficient and difficult for the victim to do the traceback.

2.1   Record route IP option[2]

The record route option provides a means to record the route of an Internet datagram. The option type is 7. A recorded route is composed of a series of Internet addresses(see *Table 1*). For the record route IP option is designed for network control use, it is seldom used in today's Internet. In the Multi-edge marking scheme, we make use of the IP option to marking the routers' IP addresses through which the packets traverse. This requires no changes to the IP protocol. Record route option is not copied on fragmentation and goes in first fragment only. It appears at most once in a datagram.

Table 1. The data format of record route IP option

0000011	length	pointer	route data

For the maximal Internet header is 60 octets, the record route IP option header is 3 octets, a typical internet header is 20 octets[2], and if there are no other IP options in use, the maximum number of IP addresses that the record route IP option can contain is $9(= (60 - 20 - 3)/4)$.

2.2   Algorithm

The multi-edge marking is to append adjacent segment attack path to the record route IP option of the packet as it travels through the network from the attacker to victim. Unlike the node append proposed in[5,16], the algorithm is a probabilistic based and does not append the routers' IP addresses to all packets. Because of the limited space in IP header, it can not mark all the routers' IP addresses into the record route IP option if the attack path is longer than 9. Here, we propose to mark the packet probabilistically. When a marking enabled router forwards a packet destined to the victim and the packet's record route IP option is not opened, it determines probabilistically whether or not to open the record route IP option of the packet to do marking. If it decides to open the option, it appends its IP address to the record route IP option. If a packet destined to the victim, the packet's record route IP option is opened and the number of the appended IP address is less than 9, it appends its IP address to the record route IP option.

After the victim collects enough marking packets, it uses the multi-edges sampled in these packets to create a graph leading back to or near to the source or sources of attack. *Figure 1* depicts the full marking and attack path reconstruction algorithms.

Marking procedure at router R:
```
for each packet w
if destination of w is victim then
    if record route IP option is set on then
        if the record route number < 9 then
            append R into w.record router IP option
        endif
    else
        let x be a random number from [0,1]
        if x < p then
            set the record route IP option on
            append R into w.record router IP option
        endif
    endif
endif
```

Path reconstruction procedure at victim v:
```
let NodeTable be an empty diagonal matrix, which element is of tuples nt(node, count)
for each path segment(R₁, R₂, ⋯ , Rₙ, 1 ≤ n ≤ 9) of attack path in the attack packet
for each router (R₁, R₂, ⋯ , Rₙ, 1 ≤ n ≤ 9)
    if Rᵢ is not in the column then
        add Rᵢ into the diagonal matrix
        set nt(Rᵢ₋₁, Rᵢ) and nt(Rᵢ, Rᵢ₊₁) be (1, 1)
    else
        increase nt(Rᵢ₋₁, Rᵢ).count, nt(Rᵢ, Rᵢ₊₁).count by 1
    endif
```
draw the attack path according to the NodeTable established. If $nt(R_i, R_j).node = 1$, R_i and R_j are connected directly

Figure 1. The multi-edge marking and path reconstruction algorithms

The marking algorithm depicts in *Figure 1* is named uncovered marking, which means that the marked IP address can not be covered by the later router.

This would lead to the further the router to victim the less probability that the router's IP address is marked. The marking algorithm can be modified to covered marking, i.e. when the IP option is full and there is another router want to mark its IP address, the very first IP address will be shifted out to give the space for marking the new comer.

If we consider the DDoS attack as propagating in a tree T, where the root of the tree T is the victim, each internal node in T corresponds to a router R on the Internet, and each leaf in T is an attack host. Our goal in the traceback problem is to identify the internal nodes of the tree T and to draw the tree. For the marked packets take the sequential segment of the path through which the packets pass, the victim need not to determine the location of each router(IP address) on the path. It just need to draw a connective graph according to the sequential segment collected. Compared with the available scheme, the victim needs not to determine which router is located before or after the other router(which may dominate the path reconstruction time), so the attack path reconstruction algorithm here is both robust and extremely quick to converge, especially in reconstructing multiple attack paths in DDoS attacks.

2.3   Analysis

The victim uses the edges marked in the attack packets to reconstruct the attack graph. The algorithm is depicted in *Figure 1*. It is noticed that the marking scheme is un-covered, which means that if a packet is marked at router A, the followed routers must mark their IP addresses into the packet unless there is no space in the record route IP option. So the probability of receiving a sample is becoming smaller the further away it is from the victim. The time for the algorithm to converge is dominated by the time to receive a sample from the furthest router. Let L be the length of the attack path, p is the marking probability and N denotes the number of the IP address that record route IP option can contain, here it is 9. The expected number of marked packet needed to reconstruct the attack path is $\lceil L/Np \rceil$. Some research results[1,19] indicate that the distance between arbitrary two hosts in Internet would not exceed 30 hops. If every marked packet carries 9 routers' address and no overlap of the path segment, 4 marked packets is enough to reconstruct the longest attack path in the Internet.

2.4   Authenticated multi-edge marking algorithm

A main disadvantage of the basic multi-edge marking scheme is that the packet markings are not authenticated. Consequently, a compromised router on the attack paths could forge the markings by appending spoofed IP address or filling up the IP record option using spoofed IP addresses, preventing the victim from determining the attack paths. To solve this problem, we need a mechanism to authenticate the packet marking. A straightforward way to authenticate the marking of the packets is to have the router digitally sign the marking. However, digital signatures are very expensive to compute and have large space overhead. Here, we propose a much efficient technique to authenticate the packet marking. The technique only uses one cryptographic MAC (Message Authentication Code) computation per marking, which is much more efficient to compute(i.e., HMAC-MD5[8,9] is three to four orders of magnitude more efficient than 1024-bit RSA signing) and can be adapted so it only requires the 16-bit overhead for storage. It is conjectured that it is computationally infeasible to produce two messages having the same message digest, or to produce any message having a given pre-specified target message digest. For the standard output of MD5 is 128-bit message, it is modified to produce a 16-bit output. In order to avoid collision, the packet-specific information is necessary. Let h_K denote the MAC function using key K. If each router R_i shares a unique secret key K_i with the victim, R_i can apply to its IP address and some packet-specific information, such as the source(S_{IP}) and destination(D_{IP}) IP addresses in the packet, with K_i, i.e. $h_{K_i}(S_{IP}, D_{IP}, R_i.IP)$ to produce the authentication $R_i.auth$. For each $R_i.auth$ is 16-bit long, the IP route record option can contain 6 marking messages at most.

The authenticated marking and attack path reconstruction algorithms are depicted in *Figure 2*.

Marking procedure at router R_i :
  for each packet w
  if destination of w is victim then
    if record route IP option is ON then
      if the record route number < 6 then
        append $R_i.IP$ into $w.record$ route IP option
        append $h_{K_i}(S_{IP}, D_{IP}, R_i)$ into $w.record$ route IP oprtion
      endif
    else
      let x be a random number from [0,1]
      if $x < p$ then
        set the record route IP option on
        append $R_i.IP$ into $w.record$ router IP option
        append $h_{K_i}(S_{IP}, D_{IP}, R_i)$ into $w.record$ route IP oprtion
      endif
    endif
  endif

Path reconstruction procedure at victim v:
  let NodeTable be an empty diagonal matrix, which element is of tuples $nt(node, count)$
  for each path segment($R_1, R_2, \cdots, R_n, 1 \leq n \leq 6$) of attack path in the attack packet
    for each router $(R_1, R_2, \cdots, R_n, 1 \leq n \leq 6)$
    if $h_{K_i}(S_{IP}, D_{IP}, R_i) = R_i.auth$ then
      if R_i is not in the column then

        add R_i into the diagonal matrix

        set $nt(R_{i-1}, R_i)$ and $nt(R_i, R_{i+1})$ be $(1,1)$

      else

        increase $nt(R_{i-1}, R_i).count$ and $nt(R_i, R_{i+1}).count$ by 1

      endif

    endif

draw the attack path according to the NodeTable established. If $nt(R_i, R_j).node = 1$, R_i and R_j are connected directly.

Figure 2.   The authenticated multi-edge marking and path reconstruction algorithms

3.   DDoS Scouter

To keep in line with the above principles, DDoS Scouter is designed as a query-respond system. Only when the IDSs deployed in the victim's system detect that there exists DDoS attacks aimed at the victim, the network administrator on receiving the marking requests asks the marking enabled routers to do IP marking. The enabled routers mark its IP address only into the specific packet(destined to the victim, the other packets destined to the other destination are not marked). The system involves the victim, intrusion detection system, network administrator or operator, IP marking and/or packet filtering enabled routers. All the communications among any components in the system must be authenticated to avoid being used by invalid users or attackers.

Figure 3 shows the architecture of the system. The DDoS Scouter consists of four entities: victim, Intrusion Detection System(IDS), network administrator and marking and filtering enabled routers.

The IDS responds to detect the DDoS attacks and sends DDoS attack alarm to the network administrators. When IDS detects that there exists DDoS attacks aimed at the victim host or network, it sends DDoS attack alarm to the victim's network administrator with the victim identity and attack characteristics. There are many commercial available IDS systems[18,21,22,23] and also there are some research results on how to detect DDoS attacks[18,21].

The network administrator is responsible for controlling the routers to do IP marking and packet filtering. On receiving the DDoS attack alarm, the network administrator authenticates that the alarm is really sent by a valid IDS. Then, it sends IP marking instructions to the IP packet marking enabled routers to start to do IP marking. On receiving the attack paths information, the network administrator decides on which routers the packet filter should be launched to stop or dilute the DDoS attacks aimed at the victim and sends the filtering instruct to the selected routers to do packet filtering. The marking and/or filtering enabled routers are responsible for carrying out the marking and packet filtering functions. On receiving the mark instructions, the IP packet

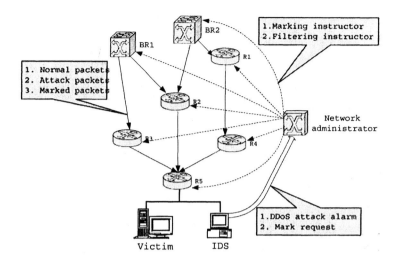

Figure 3. The architecture of the DDoS Scouter

marking enabled routers begin to mark its IP address into the packets destined to the victim. The packets routed to the other destinations are not marked. On receiving the filtering instructions, the routers filter the packets destined to the victim and forward the packets destined to the other destinations.

Having received the marked packets, the victim collects the IP addresses of routers through which the packets are passed. Using the collected the IP addresses, the victim reconstructs the attack paths or sub-paths and sends the attack paths to the network administrator to do filtering.

4. Simulation

To test the performance of the multi-edge marking scheme, we conduct an experiment on SSFnet[15], a well known network simulator system. In the simulation, the following three schemes are tested and compared: Compressed edge fragment sampling (CEFS), Un-covering Multi-Edge method (UME) and Random Multi-Edge method (RME). The first one is proposed in [5] and the last two are proposed in this paper. The difference between RME and UME lies in whether the marking procedure is random or not. In UME scheme, when the IP option is full, the following router's IP address can not be marked. In the RME scheme, when the IP option is full, the very first router's IP address is shifted out and the new router's IP address is marked in. These schemes are tested in three scenarios:

- S1: one attacker being 10 hops away from the victim.

- S2: one attacker being 24hops away from the victim.

- S3: 261 hosts locating at different places and attacking the victim through 14 different paths, i.e. a Distributed DoS attack.

Two criterias are used to make the comparison. The first one is the expected number of the packets required for path reconstruction. *Table 2* shows the simulation results. It indicates that the multi-edge marking schemes needs much less packets than CEFS, especially for the DDoS attacks. For most flooding-style DoS attacks send many hundreds or thousands of packets per second, the victim can collect the enough packet in the moment. In addition, the marking enabled router need to perform the marking function in the short time slot. From the table, we observe that when the length of the attack path increases from 10 to 24, the increment of packets in UME is slight. It indicates that UME has good adaptability for the change of the distance. The second one is the time required for path reconstruction. The simulation results indicates that all of the tests except for the CEFS in DDoS attack scenario, which takes more than one day to do the path reconstruction, can be completed within one second.

Table 2a. Number of packets needed to reconstruct the attack path $p = 0.05$

	S1	S2	S3
CEFS	2000	3600	56000
UME	40	40	800
RME	60	90	2100

Table 2b. Number of packets needed to reconstruct the attack path $p = 0.1$

	S1	S2	S3
CEFS	1400	5900	75000
UME	30	40	400
RME	70	150	2000

5.   Discussion

5.1   Fragmentation

It is indicated in[5] that the main drawback of the marking algorithm is overhead of the packet size increased by the marking, which can lead to the fragmentation and bad interactions with services such as MTU discovery[6]. Note that the routers just mark the packets destined to the victim, it does not affect the other packets in deed. If the marking results in the fragmentation of the packet, it can be designed to fragment the packet properly that the packet will be not fragmented again later. Furthermore, the fragmented packet will not be marked except for the first segment of the packet. Some research results shows that more than 95 percent attack packets are small packet, such as TCP(SYN,RST), ICMP and so on. These kind of packets have enough space for multi-edge

5.2   Authentication

In the DDoS Scouter system, there are two kinds of channels should be authenticated. The first one is the communications among the victim, the IDS, the network administrator and the marking enabled routers.The malicious users or attackers can send invalid marking request or filtering request to launch another type of DoS attacks. Any authentication and encryption mechanisms available can be deployed in the system.

For the compromised routers could forge the markings according to the precise probability distribution and preventing the victim from detecting and determining the compromised router by analyzing the marking distribution, the second one is to authenticate the marking information and has been considered in section 2.4.

5.3   Cross-domains

In DDoS Scouter, the network administrator acts as the controller to do marking and filtering on demand. Because there exists trust problem among different ISPs or ASs, the network administrator cannot send instructs to routers not belong to his domain. The attack paths reconstructed by victim is just the sub-attack paths. In order to reconstruct the full attack paths, the trust among different ISPs must be established. Based on the trust, the network administrator in the victim's domain sends the marking request and filtering request to the other ISPs network administrators. Thus, DDoS Scouter can trace back exactly to or near to the attackers or agents. A directly solution to this problem may be hierarchical mechanism.

6.   Conclusion

To make the IP traceback more practical and efficient, multi-edge marking based scheme was proposed in the paper. According to the analysis and simulation, the scheme is much more efficient than the scheme available up to now. In addition, the authors proposed a DDoS Scouter system, which is an architecture or framework, to prevent the DDoS attacks. Coupled with the fact that attack mechanisms and tools continue to improve and evolve, more effective detect-and-filter approaches must be developed in addition to the use of ingress packet filtering and other existing defense mechanisms and procedures. In the next, for the multi-edge marking scheme, we are exploring some code techniques to decrease the space requirement of one IP address. Based on the architecture, we will introduce the intelligent filtering technique into the system and extend it to a global defense infrastructure to protect the entire Internet from DDoS attacks.

References

[1] W. Theilmann, et al., "Dynamic Distance Maps of the Internet", Proc. IEEE INFOCOM'00, Tel Aviv, Israel, March, 2000.

[2] J. Postel, "Internet Protocol", RFC791, Sep. 1981.

[3] A. C. Snoeren, et al., Hash-based IP Traceback, SIGCOMM'01, August 27-31, 2001, San Diego, California, USA.

[4] Rocky K. C, Defending against Flooding-Based Distributed Denial-of-Service Attacks: A Tutorial, IEEE Communications Magazine, October 2002.

[5] S. Savage, et al., Network Support for IP Traceback, IEEE/ACM Transactions on Networking, Vol.9, No. 3, Jun. 2001.

[6] J. Mogul, et al., Path MTU discovery", RFC1191, 1990.

[7] Internet mapping, http://cm.bell-labs.com/who/ches/map/dbs/index.html, 1999.

[8] H. Krawczyk, et al., HMAC: Keyed-hashing for message authentication, Internet RFC 2104, February 1997.

[9] R. L. Rivest, The MD5 message digest algorithm, RFC 1321, Internet Activities Board, Internet Privacy Task Force, April 1992, 1992.

[10] Steve Bellovin, The icmp traceback message, http://www.research.att.com/?smb, 2000.

[11] P. Ferguson, et al., Network ingress filtering: Defeating denial of service attacks which employ ip source address spoofing, RFC 2267, January 1998.

[12] Hal Burch, et al., Tracing anonymous packets to their approximate source", Unpublished paper, December 1999.

[13] Robert Stone, Centertrack: An ip overlay network for tracking dos floods, Unpublished, October 1999.

[14] Drew Dean, et al., An algebraic approach to ip traceback, in Network and Distributed System Security Symposium, NDSS '01, February 2001.

[15] SSFnet, http://www.ssfnet.org

[16] D. X. Song, et al., Advanced and Authenticated Marking Schemes for IP Traceback, Proc. IEEE INFOCOM'01, April, 2001.

[17] D. Moore, et al., Inferring Internet Denial-of-Service Activity, Proc. Of the 10th USENIX Security Symposium, Washington, D.C., USA, August, 2001.

[18] H. Wang, et al., Detecting SYN Flooding Attacks", Proc. IEEE INFOCOM'02, 2002.

[19] R. L. Carter, et al., Dynamic Server Selection Using Dynamic Path Characterization in Wide-Area Networks, Proc. IEEE INFOCOM'97, Kobe, Japan, April, 1997.

[20] M. T. Goodrich, Efficient Packet Marking for Large-Scale IP Traceback, CCS'02, November 18-22, 2002, Washington, DC, USA.

[21] A. Ramanathan, et al., WADeS: A Tool for Distributed Dennial of Service Attack Detection, ACM SIGCOMM Internet Measurement Workshop 2002.

[22] C. Manikopoulo, et al., Network Intrusion and Fault Detection: A Statistical Anomaly Approach, IEEE Communications Magazine, October, 2002.

[23] G. Vigna, et al., NetSTAT: A Network-based Intrusion Detection Approach, Proc. 14th An. Comp. Sec. App. Conf., 1998.

A METHOD OF DIGITAL DATA TRANSFORMATION–BASE91

He Dake

School of Computer and Communication Engineering
Southwest Jiaotong University, Chengdu 610031, P. R. China
hedk@home.swjtu.edu.cn

He Wei

Scientific Research Office
Southwest Jiaotong University, Chengdu 610031, P. R. China
heweikiko870@sina.com.cn

Abstract   A coding transformation method, called Base91, is characterized by its output of 91 printable ASCII characters. Base91 provides compatibility with the E-mails and increases the encoding efficiency of input enciphered E-mail's data or any input 8-bit data sequence. Its extension Base91+ has a higher encoding efficiency.

Keywords: coding transformation, Base91, Base85, Base64, QP, encoding efficiency

1.   Background of Invention

With the rapid development of Internet and its business application, E-mail and its security has become more and more important. SMTP (Simple Mail Transfer Protocol) is the basic electronic mail transfer protocol. All the SMTP-based E-mail encrypting system PGP (Pretty Good Privacy), PEM (Privacy Enhanced Mail), and MIME (Multipurpose Internet Mail Extensions) or S/MIME (secure MIME) can provide compatibility with the E-mails. So-called compatibility with the E-mails is to transform arbitrary 8-bit data byte-strings or arbitrary bit stream data transferred by the E-mail into a character-strings of a limited ASCII (American Standard Code for Information Interchange). The main limitation on the latter is that: (a) the characters have to be printable; (b) the characters are not control character or "-"(hyphen). There are totally 94 of such ASCII characters, their corresponding digital coding being all integers ranging from 32 through 126 with the exception of 45. E-mails written in these

ASCII characters are compatible with the Internet standard SMTP, and can be transferred in nearly all the E-mail systems. Nowadays, to provide compatibility with the E-mail, Base64 coding or QP (Quoted-Printable) coding is usually employed.

Base64 coding divides the input message into blocks 6-bit long to be used as variable implementation mapping, the mapping is denoted by

Base64[ ]: $X \rightarrow Y$

where the variable or original image set X includes all 64 6-bit long symbols (denoted as integers 0, 1, ..., 63) and representing "no data"; the image set Y includes the upper and lower cases of 26 alphabetic characters, Arabic digits ranging from 0 through 9, "+", "/" and filling character "=", where it is specified that in the non-program statements the Chinese quotation marks are used as the delimiter of characters or character-strings (the following is the same). Mapping rules commonly used in Base64 coding software are

$$Base64[0]="A",\ldots, Base64[25]="Z",$$
$$Base64[26]="a",\ldots, Base64[51]="z",$$
$$Base64[52]="0",\ldots, Base64[61]="9",$$
$$Base64[62]="+",\quad Base64[63]="/"$$

Particularly, Base64[Φ]="=" is used only when needed so as to make the total number of characters of output string equal to the multiples of 4, where Φ being empty set. The coding efficiency of Base64 coding is 6/8 = 75%. The data expansion rate is 8/6 = 4/3 = 133.33%.

QP coding divides the input message into blocks 8-bit long to be used as variable implementation mapping, when the original image 8-bit data is non-"=" printable character, its image equal to the original image (i.e. there is no change); when the hexagonal notation of the original image 8-bit data is "LR" and the most significant bit is 1, its image is three printable characters "=LR"; while the image of "=" is "=3D". Hence, in the worst case, the encoding efficiency of QP transformation is 1/3 and the data expansion rate is 300%, (it is the case that Chinese language data employing coding GB2312 are being QP-transformed).

Base85 coding transforms an input data of 16 bytes into an output data of 20 printable ASCII characters, so the encoding efficiency of Base85 is 16/20 or 80%. But people only use Base85 in the representation of 128-bit address of IPv6 now.

2.   Contents of Invention

The object of the present invention is to provide a digital data transformation method to replace Base64 coding or QP coding, so as to provide higher coding efficiency under the condition of E-mail compatibility, to reduce the time requirement for transferring coding messages over the network, or to save storage space when the data are stored using printable character mode.

The present invention will be implemented by the following technical design: a coding transformation of arbitrary bit stream data into printable character sequence. The main idea is: to increase the bit length of the block mapping from the current 6 or 8 bits to 13 bits , and to use the double-character set of 91 printable ASCII characters as the image set for transformation. The followings are the Base91 coding designed for the present invention (also denoted as Radix-91 coding, where Base91 and Radix-91 are two short names of "base number-91"). Then, Base91+ coding with a block of 27 bits is given as an extension of Base91.

2.1   Base91 Coding

Base91 coding divides the input message into blocks 13-bit long to be used as variable implementation mapping, the mapping is denoted by

Base91[ ]: X →Y

where the variable or original image set X includes all 8192 13-bit long symbols (denoted as integers $0,1,\ldots,8191$) and, symbols $\phi_n=8191+n$ $(n=1,\ldots,12)$, denoting that the n-bit data at the specified side of the last block are used as the filling data, thereby making the total number of elements in the original image set equal to 8204; the image set Y is the sub-set of the direct product of R91×R91, where the symbol R91 denotes the set of 91 characters selected from the 95 printable ASCII character set with " - ","= ",". " and space characters excluded, the direct product R91×R91 has 8281 elements.

Base91 is defined as an injective mapping arbitrarily selected from X into the direct product R91×R91. The selection of any particular injective mapping as Base91 has no effect on the present invention. For the convenience of implementation, assuming that R91_CH[91] is the character set that includes all R91 characters and is arranged according to the ASCII sequential order, the present invention preferably selects the following mapping :

$$Base91[x] = (ch1, ch2) = (R91\_CH[x/91], R91\_CH[x\%91]) \quad (1)$$

where x∈X, ch1,ch2∈R91, symbols "/" and "%" are the operators used in the C language,representing integral division and modulo division (remainder) respectively.

The operation of dividing the input message into 13-bit long blocks may produce the last block less than 13-bit long. For such blocks, n bits are added to the specified side to make it become a complete block for implementing mapping; and a block of data ϕ_n is added thereafter as the input data implementing mapping so that it can be decided how many filling bits have to be deleted during decoding. When needed, double-character "==" may be used as a "terminating symbol" of the output character-string. Hence at most 92 printable ASCII characters can appear in the output-string of Base91 coding.

According to the coding rules of the above-mentioned Base91 coding, the number of extra added output data consisting of the filling bits, the image of the denoting symbols and the "terminating symbol" does not exceed 6 characters. Therefore, with the increase of the bit number or byte number of the input message, the coding efficiency of the Base91 approaches 81.25%, its data expansion rate approaches 123%.

2.2   Base91+ Coding

Base91+ coding divides the input message into blocks 27-bit long to be used as variable implementation mapping, the mapping is denoted by

Base91+[ ]: $X \rightarrow Y$

where the variable or original image set X includes all 134217728 27-bit long symbols (denoted as integers $0, 1, \ldots, 134217727$) and symbols $\psi_n = 134217727 + n$ $(n = 1, \ldots, 26)$, denoting that the n-bit data at the specified side of the last block are used as the filling data, thereby making the total number of elements in the original image set equal to 134217754; the image set Y is the sub-set of the direct product of $Y0 \times Y0$, where the symbol Y0 is a sum set of $R91 \times R91$ and HZm[],which is a subset of GB2312 and with m elements, $m = 3305$. That is

Y0= { R91×R91 } $\cup$ HZm[3305]

N=8281+m=11586, the number of Y0, is called "extended base number". $N^2 > |X| = 134217754$, that is,the direct product $Y0 \times Y0$ has more elements than X.

Base91+ is defined as an injective mapping arbitrarily selected from X into the direct product $Y0 \times Y0$. The selection of any particular injective mapping as Base91+ has no effect on the present invention. For the convenience of implementation, the present invention preferably selects the following mapping :

$$Base91 + [x] = (ch1, ch2, ch3, ch4), \quad x \in X, \tag{2}$$

where, by the help of y1=x/N and y2=x% N,
if $y1 < 8281$

$$ch1 = R91\_CH[y1/91], ch2 = R91\_CH[y1\%91] \tag{3}$$

if $y1 \geq 8281$
$$ch1ch2 = HZm[y1 - 8281] \tag{4}$$

if $y2 < 8281$
$$ch3 = R91\_CH[y2/91], ch4 = R91\_CH[y2\%91] \tag{5}$$

if y2≥8281

$$ch3ch4 = HZm[y2 - 8281] \tag{6}$$

The operation of dividing the input message into 27-bit long blocks may produce the last block less than 27-bit long. For such blocks, n bits are added to the specified side to make it become a complete block for implementing mapping; and a block of data ψ_n is added thereafter as the input data implementing mapping so that it can be decided how many filling bits have to be deleted during decoding.

Compared with the Base64 or QP coding, the Base91 (or Base91+) has its advantage in encoding efficiency. The design features of the four kinds of coding transformation are shown in Table 1.

Table 1.

design features	QP coding MSB of input byte being 1	Base64 coding	Base91 coding	Base91+ coding
number of bits of one input block	8	6	13	27
number of elements of variable set X	256	64	$2^{13}+12$	$2^{27}+26$
number of bits using by one image element	24	8	16	32
number of characters in output	17	65	91	91+94
encoding efficiency	33.333%	75%	81.25%	84.375%
data expansion rate	300%	133%	123%	118.5%
ratio of amount of coded data of same input message	225	100	92.3	88.89
	100	44.44	41.03	39.51

3.   Conclusion

Base91 provides compatibility with the E-mails and increases the encoding efficiency of input enciphered E-mail's data or any input 8-bit data sequence. Combined with Internet standards SMTP, MIME, S/MIME etc., Base91 encoding can reduce 7.7% of transmitted data required by Base64 encoding, and can reduce 58.97% of transmitted data required by QP encoding with MSB of every input byte being 1 as in the input data of Chinese GB2312, which is a subset of GBK. Its extension Base91+ has a higher encoding efficiency of 84.375%, which is high with 9.375 percent. than 75% of Base64. In other words, Base91+ encoding can reduce 11.11% of transmitted data required by Base64 encoding and can reduce 60.49% of transmitted data required by QP encoding with MSB of every input byte being 1.

References

[1] A digital data transforming method , P.R.China Patent Application No.00112884.1, 2000. 04. 28, Publicity date 2000.10.11, inventors:He Dake, He Wei.

[2] A digital data transforming method, PCT/CN01/00615, 2001.04.26, Publicity No. WO 02/33828A1, 2002.04.25,inventors:He Dake, He Wei.

[3] A method of digital data transformation, U.S.Patent Application No.10/240.707, October 3,2002,Applicants and inventors: He Dake, He Wei.

[4] A method of digital data transformation,European Patent Application No.01937948.6-2206-CN0100615, Dec.02.2003,inventors:He Dake, He Wei.

AN APPROACH TO THE FORMAL ANALYSIS OF TMN PROTOCOL *

ZHANG Yu-Qing, LIU Xiu-Ying

National Computer Network Intrusion Protection Center, GSCAS, Beijing 100039, China
zhangyq@mail.nipc.org.cn, liuxy@mail.nipc.org.cn

Abstract   This paper analyzes the TMN protocol completely using a formal analysis method called the Running-Mode Analysis and uncovers a number of attacks on the TMN protocol. These attacks are classified according to the detailed forms and the different intentions of the intruder. Finally, combining with the known attacks, the authors deduce that the Running-Mode Analysis can analyze the TMN protocol effectively.

Keywords: TMN protocol, model checking, cryptographic protocol, running-mode analysis

1.   Introduction

TMN protocol [1] due to Tatebayashi, Matsuzaki and Newman concerns a mobile communications system. In order for two agents to set up a secure session, communicating over an open channel, they must first decide upon a cryptographic session key, which should be kept secret from all the eavesdroppers. The protocol is subject to a number of attacks. Two attacks are presented by Murphi in reference [2]. Reference [3] uncovers two attacks on TMN protocol. Reference [4] analyzes completely the TMN protocol by using model checking tool FDR to uncover seven attacks on the original protocol and three attacks on the fixed attacks.

We use a new formal analysis method called the Running-Mode method [5] to analyze the TMN protocol. Nineteen attacks in the small system are found. Combining with the known attacks, we notice that many attacks on the TMN protocol are similar. So there are repetitious attacks presented. In this paper, we classify these attacks combining with the secret aim of the protocol.

Section 2 presents an introduction to the TMN protocol, and section 3 presents an analysis of the protocol by using the Running-Mode method. In section 4, the

*This work is supported by National Natural Science Foundation of China under Grant 60102004,60273027,60025205

attacks on the TMN protocol are classified into different kinds. A conclusion is drawn in section 5.

2.   The TMN protocol

The TMN protocol concerns three principals: an initiator A, a responder B, and a server S who mediates between them as follows:

M1 $A \rightarrow S : B, \{N_a\}_{Ks}$
M2 $S \rightarrow B : A$
M3 $B \rightarrow S : A, \{N_b\}_{Ks}$
M4 $S \rightarrow A : B, \{N_b\}_{Na}$

Here, A is an initiator, B is a responder, S is a server. Ks is the public key of the server, N_a and N_b are the nonces of A and B. In order to establish the secret with B, A must send a message to S, and he must inform S communicating with B and send the nonce N_a encrypted by the S' public key (message M1). After receiving A's message, S contacts with B (message M2). Then B accepts the A's request and send N_b to S (message M3). S encrypts the nonces of N_a and N_b and sends them to A. Finally A gets the shared secret N_b (message M4). The protocol employs two sorts of encryption:

Standard encryption: This uses an encryption function, which we shall write as E. Every initiator and responder know how to produce E(m) given message m, but only the server knows how to decrypt such a message to obtain the original message m. This encryption can be implemented using, for example, RSA. Message 1 and Message 3 use this encryption.

Vernam encryption: The Vernam encryption of two keys, which we will write as $V(k_1, k_2)$, is their bit-wise exclusive-or. Note that $V(k_1, V(k_1, k_2))=k_2$, so if an agent knows k_1, then he can decrypt $V(k_1, k_2)$ to obtain k_2. Message 4 uses this encryption.

3.   Analysis of TMN protocol using Running-Mode

3.1   Introduction to the Running-Mode method

The basic method of model checking is to produce a model of a small system running the protocol, together with a model of the most general intruder who can interact with the protocol, and to use a state exploration tool to discover if the system can enter an insecure state, that is, whether there is an attack upon the protocol. An approach of Running-Mode analysis is deduced from some results for model checking of security protocols. Therefore, the basic approach of Running-Mode analysis is also to produce a model of a small system running the protocol, together with a model of the most general intruder who can interact with the protocol, and to analyze all the possible running modes of this system.

In our system, the intruder is an unhonest principals:

(1) Overhear and/or intercept any messages being passed in the system;

(2) Decrypt messages that are encrypted with his public key so as to learn new nonces;

(3) Introduce new messages into the system, using nonces he knows;

(4) Replay any message he has seen (possibly changing plain-text parts), even if he does not understand the contents of the encrypted part.

3.2 Analysis of the TMN protocol

3.2.1 The model of the small system. We define a small system, which has an intruder. The system configuration is one initiator A, one responder B, one server S and one intruder I.

The sets of the system are as follows,Σ denotes the small system: $\Sigma = \{$ID, Key,Message$\}$; ID=$\{$InitSet,RespSet,ServSet$\}$,ID denotes the set of principals. The set of initiators is InitSet=$\{A, I, I(A), \Phi\}$ Here A is an honest initiator who can run the protocol precisely once; I is an intruder; $I(A)$ denotes that I impersonates A. Φ denotes that there is not any principal. The set of responders is RespSet=$\{B, I, I(B), \Phi\}$. Here B is an honest responder who can run the protocol precisely once; I is an intruder; $I(B)$ denotes that I impersonates B. The set of servers is ServSet=$\{S, I(S), \Phi\}$. Here S is a trusty principal; $I(S)$ denotes that I impersonates S. Key=$\{Ks, N_a, N_b, N_i\}$ denotes the set of keys and nonces. Message=$\{$M1,M2,...,Mn$\}$ denotes the set of messages.

3.2.2 The modes of The TMN protocol. Model checker can automatically analyze the concurrent protocol runs. To verify that a protocol is correct, all the possible runs must be checked. But the method of the Running-Mode can not analyze automatically like model checker and we must discuss the concurrent protocol runs before using it.

When a protocol runs concurrently, it must satisfy the assumptions on the small system. Here the honest initiator A and the honest responder B can run the protocol precisely, they might lead to two runs of the protocol. Moreover the intruder I can run the protocol with the server by different impersonates. If the information is transmitted between the intruder and the impersonators, the intruder can not get the beneficial information. We can consider these runs as once. Therefore, we can make a conclusion that the concurrent three-principal cryptographic protocol runs is no more than three times.

3.2.2.1 The modes when the TMN protocol runs only once. When the TMN protocol runs only once, the running mode is as follows:

1.1 $X_1 \rightarrow Z_1 : Y, \{N_{X1}\}_{Ks}$

1.2 $Z_2 \rightarrow Y_1 : X$

1.3 $Y_2 \rightarrow Z_3 : X, \{N_{Y2}\}_{Ks}$

1.4 $Z_4 \to X_2 : Y, \{N_{Y2}\}_{N_{x1}}$

Here, $X, X_1, X_2 \in \{A, I, I(A), \Phi\}; Y, Y_1, Y_2 \in \{B, I, I(B), \Phi\}; Z_i \in \{S, I(S), \Phi\}$, $i = 1, \ldots, 4; N_{X1} \in \{N_a, N_i\}; N_{Y2} \in \{N_b, N_i\}$.

3.2.2.2 The modes when the TMN protocol runs concurrently.

We know that the concurrent run of the protocol means that the protocol runs several times at a time, i.e. the protocol runs several times before the first run ends.

(1) When the TMN protocol runs two times concurrently, the running mode is as follows£°

1.1 $X_1 \to Z_1 : Y, \{N_\alpha\}_{Ks}$

1.2 $Z_2 \to Y_1 : X$

1.3 $Y_2 \to Z_3 : X, \{N_\beta\}_{Ks}$

1.4 $Z_4 \to X_2 : Y, \{N_\beta\}_{N\alpha}$

2.1 $X_3 \to Z_5 : Y', \{N_\alpha'\}_{Ks}$

2.2 $Z_6 \to Y_3 : X'$

2.3 $Y_4 \to Z_7 : X', \{N_\beta'\}_{Ks}$

2.4 $Z_8 \to X_4 : Y', \{N_\beta'\}_{N\alpha'}$

Here, $X, X', X_i \in \{A, I, I(A), \Phi\}, i = 1, \ldots, 4; Y, Y', Y_i \in \{B, I, I(B), \Phi\}, i = 1, \ldots, 4; Z_j \in \{S, I(S), \Phi\}, j = 1, \ldots, 8; N_\alpha, N_\alpha' \in \{N_a, N_i\}; N_\beta, N_\beta' \in \{N_b, N_i\}$.

In the small system, the honest principals A and B can run the protocol only once. Therefore, the variables must be satisfied with the following conditions:

$X_1 \oplus X_2 = A, X_3 \otimes X_4 \neq A; Y_1 \oplus Y_2 = B, Y_3 \otimes Y_4 \neq B;$

$X_3 \oplus X_4 = A, X_1 \otimes X_2 \neq A; Y_3 \oplus Y_4 = B, Y_1 \otimes Y_2 \neq B.$

Here the symbol $\oplus$ denotes OR, $\otimes$ denotes AND.

(2) When the TMN protocol run three times concurrently, the mode is as follows£°

1.1 $X_1 \to Z_1 : Y, \{N_\alpha\}_{Ks}$

1.2 $Z_2 \to Y_1 : X$

1.3 $Y_2 \to Z_3 : X, \{N_\beta\}_{Ks}$

1.4 $Z_4 \to X_2 : Y, \{N_\beta\}_{N\alpha}$

2.1 $X_3 \to Z_5 : Y', \{N_\alpha'\}_{Ks'}$

2.2 $Z_6 \to Y_3 : X'$

2.3 $Y_4 \to Z_7 : X', \{N_\beta'\}_{Ks}$

2.4 $Z_8 \to X_4 : Y', \{N_\beta'\}_{N\alpha'}$

3.1 $X_5 \to Z_9 : Y'', \{N_\alpha''\}_{Ks}$

3.2 $Z_{10} \to Y_5 : X''$

3.3 $Y_6 \to Z_{11} : X'', \{N_\beta''\}_{Ks}$

3.4 $Z_{12} \to X_6 : Y'', \{N_\beta''\}_{N\alpha''}$

Here, $X, X', X'', X_i \in \{A, I, I(A), \Phi\}, i = 1, \ldots, 6; Y, Y', Y'', Y_i \in \{B, I, I(B), \Phi\}, i = 1, \ldots, 6; Z_j \in \{S, I(S), \Phi\}, j = 1, \ldots, 12; N_\alpha, N_\alpha', N_\alpha'' \in \{N_a, N_i\}; N_\beta, N_\beta', N_\beta'' \in \{N_b, N_i\}$.

The variables must be satisfied with the following conditions£°

$$X_1 \oplus X_2 = A, X_3 \otimes X_4 \otimes X_5 \otimes X_6 \neq A;$$
$$Y_1 \oplus Y_2 = B, Y_3 \otimes Y_4 \otimes Y_5 \otimes Y_6 \neq B;$$
$$X_3 \oplus X_4 = A, X_1 \otimes X_2 \otimes X_5 \otimes X_6 \neq A;$$
$$Y_3 \oplus Y_4 = B, Y_1 \otimes Y_2 \otimes Y_5 \otimes Y_6 \neq B;$$
$$X_5 \oplus X_6 = A, X_1 \otimes X_2 \otimes X_3 \otimes X_4 \neq A;$$
$$Y_5 \oplus Y_6 = B, Y_1 \otimes Y_2 \otimes Y_3 \otimes Y_4 \neq B.$$

Then the Running-Mode of the protocol is to give the different values to the different variables, now we can list all the running modes of the TMN protocol.

3.3   Reduction of the number of the running modes

When using the Running-Mode method to analyze some protocols, we obtain all the possible running modes by giving the different values to the different variables and then analyze the modes to find out whether there exits any attack. In order to reduce the work by hand, we can reduce some impossible modes by the following rules:

(1) If the value of the messages' sender or receiver in the protocol is Φ, this message is invalid and we do not need to consider this instance;

(2) If the secret information in the message does not match the identity of its sender, for example, the honest principal A or B sending the nonce N_i, we do not need to consider this instance because it is impossible in the real run of the protocol;

(3) If all the participants in the protocol are honest principals, the protocol runs normally and does not lead to any attack, we do not need to consider this instance;

(4) If both the initiator and the responder are the different impersonates of the intruder, which means that the information transfers between the different identities of the intruder and does not lead to any attack, we do not need to consider this instance;

(5) In the concurrent runs of the protocol, if every run has not any impact on the other, i.e. in every run of the protocol the information transferred in other runs is not used, something like the independent run, we do not need to analyze this instance.

4.   Attacks on the TMN protocol

4.1   The attacks when the protocol runs only once

When we replace all the variables with all the possible values in the only one run of the TMN protocol, we uncover six attacks.

$(1) X_1 = X_2 = X = A, Y_1 = Y_2 = I(B), Y = B, Z_i = S(i=1,\ldots,4), N_{X1} = N_a,$ $N_{Y2} = N_i;$

(2)$X_1=X_2=X=A,Y_1=Y=B,Y_2 = I(B),Z_i = S(i=1,\ldots,4),N_{X1} = N_a,$
$N_{Y2} = N_i;$
(3)$X_1=X_2 = I(A),X=A,Y_1=Y_2=Y=B,Z_i = S(i=1,\ldots,4),N_{X1} = N_i,$
$N_{Y2} = N_b;$
(4)$X_1=X=A,X_2 = I(A),Y_1= Y_2 = I(B),Y=B,Z_i = S(i=1,\ldots,4),$
$N_{X1} = N_a, N_{Y2} = N_i;$
(5)$X_1=X=A,X_2 = I(A),Y_1=Y=B,Y_2 = I(B),Z_i = S(i=1,\ldots,4),$
$N_{X1} = N_a,N_{Y2} = N_i;$
(6)$X_1=X_2=Z_1=Z_4=\Phi,X=A,Y_1=Y_2 = B,Y=\Phi, Z_2=Z_3 = I(S),$
$N_{Y2} = N_b,N_{X1}$=anything.

In both attack 1 [2] and attack 2 [3], the intruder I sends his nonce N_i to deceive A by impersonating the responder B, thus making A think that he has a shared secret information with B, but I decrypts the Vernam function to get the secret information N_a. In fact, both attacks reach the same goal by the stay-in-mid attack, then we classify these two into the first kind of attack.

In attack 3 [2], the intruder I gets the secret information N_b by impersonating the initiator A and deceives the honest responder B, although it is also the stay-in-mid attack, the goal is different from that of the first kind of attack, then we classify it into the second kind.

In attacks 4 and 5, the intruder impersonates both A and B to take part in the run of the protocol and he gets the secret information N_a. Then the intruder can use this secret information to make the attack in the other run of the protocol, so we classify them into the third kind.

In attack 6 [2], the intruder deceives the honest responder B by impersonating the server S, the intruder achieves this goal because of the flaw that the protocol does not verify the identity of the initiator A. Although the main goal of the TMN protocol is not to verify the identities of the communicators, our method still can find this leak of this protocol, then we classify this attack into the fourth kind.

4.2   The attacks when the protocol runs concurrently

When the protocol runs concurrently, we place the possible values to the variables in the running modes and use our remove rules in section 3.3 to reduce the impossible modes, then we obtain the following attacks:

(1) $Y_1=Y_2=Z_2=Z_3=X=\Phi,Z_5=Z_6=Z_7=Z_8 = S,Z_1=Z_4 = I(S)$,
$N_\alpha=N_\alpha'=N_a, N_\beta=N_\beta'=N_i,X_1=X_2 = A,X_3=X_4=X'=I,Y_3=Y_4 = I(B),$
$Y=Y'=B;$
(2) $Y_1=Y_2=Z_2=Z_3=X=\Phi,Z_5=Z_6=Z_7=Z_8 = S,Z_1=Z_4 = I(S),$
$N_\alpha=N_\alpha'=N_a,N_\beta=N_\beta'=N_i, X_1=X_2=X'=A,X_3=X_4 = I(A),$
$Y_3=Y_4 = I(B),Y=Y'=B;$

(3) $Y_1=Y_2=Z_2=Z_3=X=\Phi,Z_5=Z_6=Z_7=Z_8 = S,Z_1=Z_4 = I(S)$,
$N_\alpha=N_\alpha'=N_a,N_\beta=N_\beta'=N_i$, $X_1=X_2=X'=A,X_3=X_4 = I(A)$,
$Y_3=Y_4=Y'=I,Y=B$;

(4) $Y_1=Y_2=Z_2=Z_3=X=\Phi,Z_5=Z_6=Z_7=Z_8 = S,Z_1=Z_4 = I(S)$,
$N_\alpha=N_\alpha'=N_a,N_\beta=N_\beta'=N_i$, $X_1=X_2 = A,X_3=X_4=X'=Y_3=Y_4=Y'=I$,
$Y=B$;

(5) $X_3=X_4=Z_5=Z_8=Y'=\Phi,Z_1=Z_2=Z_3=Z_4 = S,Z_6=Z_7 = I(S)$,
$N_\alpha=N_\alpha'=N_i,N_\beta=N_\beta'=N_b,X_1=X_2 = I(A),X=X'=A,Y_1=Y_2 = I(B)$,
$Y_3=Y_4=Y=B$;

(6) $X_3=X_4=Z_5=Z_8=Y'=\Phi,Z_1=Z_2=Z_3=Z_4 = S,Z_6=Z_7 = I(S)$,
$N_\alpha=N_\alpha'=N_i,N_\beta=N_\beta'=N_b,X_1=X_2 = I(A),X=X'=A,Y_1=Y_2=Y=I$,
$Y_3=Y_4 = B$;

(7) $X_3=X_4=Z_5=Z_8=Y'=\Phi,Z_1=Z_2=Z_3=Z_4 = S,Z_6=Z_7 = I(S)$,
$N_\alpha=N_\alpha'=N_i,N_\beta=N_\beta'=N_b,X_1=X_2=X=I,X'=A,Y_1=Y_2 = I(B)$,
$Y_3=Y_4= Y=B$;

(8) $X_3=X_4=Z_5=Z_8=Y'=\Phi,Z_1=Z_2=Z_3=Z_4 = S,Z_6=Z_7 = I(S)$,
$N_\alpha=N_\alpha'=N_i,N_\beta=N_\beta'=N_b,X_1=X_2=X=Y_1=Y_2=Y=I,X'=A$,
$Y_3=Y_4 = B$;

(9) $Z_1=Z_2=Z_3=Z_4=Z_5=Z_6=Z_7 = S,Z_8=X_4=\Phi$,
$X_1=X_2=X=Y_1=Y_2=Y=I,X_3=X'=A,Y_3=Y_4=Y'=B$,
$N_\alpha = N_i,N_\alpha'=N_a,N_\beta=N_\beta'=N_b$;

(10) $Z_1=Z_2=Z_3=Z_4=Z_5=Z_6=Z_7=Z_8 = S,X_1=X_2=X=A$,
$X_3=X_4=X'=I,Y_1=Y_2=Y=Y'=B,Y_3=Y_4 = I(B),N_\alpha = N_i$,
$N_\alpha'=N_a,N_\beta=N_\beta'=N_b$;

(11) $Z_1=Z_2=Z_3=Z_4=Z_5=Z_6=Z_7=Z_8 = S,X_1=X_2=X=X'=A$,
$X_3=X_4 = I(A),Y_1=Y_2=Y=B,Y_3=Y_4=Y'=I,N_\alpha= N_\alpha'=N_a$,
$N_\beta = N_b,N_\beta'=N_i$;

(12) $Z_1=Z_2=Z_3=Z_4=Z_5=Z_6=Z_7=Z_8 = S,X_1=X_2 = I(A)$,
$X_3=X_4=X=X'=A,Y_1=Y_2=Y=Y'=B,Y_3=Y_4 = I(B),N_\alpha = N_i$,
$N_\alpha'=N_a,N_\beta=N_\beta'=N_b$;

(13) $Y_1=Y_2=Z_2=Z_3=X=\Phi,Z_5=Z_6=Z_7=Z_8 = S ,Z_1=Z_4 = I(S)$,
$N_\alpha=N_\alpha'=N_a,N_\beta=N_\beta'=N_b,X_1=X_2 = A,X_3=X_4=X'=I$,
$Y_3=Y_4=Y=Y'=B$.

Now we analyze the attacks in the concurrent run of the protocol.
Attack 1: the first run of the protocol

1.1 $A \rightarrow I(S):B,\{N_a\}_{Ks}$

now begins the second run of the protocol

2.1 $I \rightarrow S:B,\{N_a\}_{Ks}$

2.2 $S \rightarrow I(B):I$

2.3 $I(B) \rightarrow S:I,\{N_i\}_{Ks}$

2.4 $S \rightarrow I:B,\{N_i\}_{Na}$

the first run continues

1.4 $I(S) \rightarrow A{:}B, \{N_i\}_{Na}$

In the first run of the protocol, I eavesdrops message 1.1 sent from A to the server S in the first step. Because I does not know the private key of the server S, he can not decrypt this information. Then in the second step, the intruder I runs another run of the protocol by its own identity, he sends message 2.1 to the server S to make commutation with B and replays message 1.1(eavesdropped before) to S, then he intercepts and captures message 2.2 sent from S to B by impersonating B, at the same time, he sends its nonce N_i to S (message 2.3). At the end of the second run of the protocol, I can get N_a. Finally, I replays message 2.4 to A by impersonating S and makes A think that he has the session key N_i shared with B (message 1.4), but in fact, this key is the shared key between A and I.

The scenarios of the attacks 2,3 and 4 are almost the same as attack 1, the essence of them is that the intruder replays message 1.1 by having the second run of the protocol, and he decrypts the secret information to get N_a transferred in the first run by impersonating different identities, then I replays message 2.4 by impersonating the server, which makes A think that he has get the shared key with B but A is deceived. In a word, these attacks obtain secret message N_a and deceive A by replay attack, then we classify them into the fifth kind of attack.

The essence of the attacks 5, 6, 7 [2] and 8 [4] is that in the first run of the protocol, the intruder takes part in the run by different identities. After message 1.2, I has the second run of the protocol by impersonating the server S and deceives B to send a shared key N_b with A. Then the first run of the protocol continues and the intruder replays the message 2.3(from the second run) and he can get N_b. In a word, these attacks get the secret information N_b and deceive B by replay attack, then we classify them into the sixth kind of attack.

Attack 9 [3] and attack 10 [2] belong to the same kind. Because in these attacks, the intruder listens in the messages in the formal run of the protocol, and he requests the server to have communication with himself or B by his own identity in the second run of the protocol (if he wants to have communication with B, he intercepts and captures the messages sent from S to B ; if he wants to have communication with himself, he replays message 1.3 directly.)Finally, the intruder can get N_b , but he does not deceive any honest participant. Then we classify them into the seventh kind of attack.

In attack 11 [2], the intruder listens in the messages in the formal run of the protocol, then he requests the server to have communication with himself by impersonating A and replays message 1.1, finally, I gets N_a. Then we classify this attack to the eighth kind of attack.

In attack 12 [2], the intruder puts the two attacks in the independent runs of the protocol together and forms the new attack. The intruder can get N_b in the first run of the protocol and use N_b to get N_a in the second run. Then we classify this attack into the ninth kind of attack.

In attack 13, the intruder receives message 1.1 from A by impersonating S, then I begins the second run of the protocol. In message 2.1, the intruder replays message 1.1, but he can not decrypt message 2.4 because he does not know the secret information N_a in this message, then he can only replay message 2.4 in the message 1.4 and make A think that he has established the shared key with B, then A is deceived. This attack uncovers the leak in verifying the identity of B, and we classify it into the tenth kind of attack.

By now, we have not uncovered any attack in the TMN when it runs three times concurrently. However, we have made a complete analysis of the TMN protocol using our Running-Mode method and uncovered nineteen attacks which are classified into ten kinds.

5.   Conclusion

We have analyzed the TMN protocol using the Running-Mode method and got nineteen attacks on the TMN protocol. In our small system, we assure that our analysis is complete. This method can not only verify the result of model checking, but also uncover new attacks or weakness. Therefore, the Running-Mode method is an effective method of the cryptographic protocol analysis.

References

[1] M.Tatebayashi,N.Matsuzaki,and D.B.Newman.Key distribution protocol for digital mobile communication systems.In advance in cryptology——CRYPTO'89,volume 435 of LNCS,324-333.Springer-Verlag,1989.

[2] J.C.Mitchell,M.Mitchell and U.Stern,Automated Analysis of Cryptographic Protocols using Murphi,in Proceeding of the IEEE Symposium on Security and Privacy,May 1997,141-151.

[3] Yuqing ZHANG,Yupu HU,Guozhen XIAO. Some new attacks on the TMN cryptographic protocol.Journal of XIDIAN University,2000,27(1):130-132.

[4] G.Lowe and A.W.Roscoe.Using CSP to detect errors in the TMN protocol. Software Engineering, 1997,23(10):659-669.

[5] Yuqing ZHANG,Xiuying LIU. An approach to formal verification of the three-principal cryptographic protocols. ACM Operating Systems Review, 2004, 38(1): 35-42.

Index

access authentication, 181
attack, 165
authenticated key agreement protocol, 151
authentication, 157
authentication codes, 33

Base64, 229
Base85, 229
Base91, 229
bilinear pairings, 59
binary expression of integer sums, 45
blind signature, 73, 123, 129
block cipher, 173
broadcasting, 165

CHAP, 181
coding transformation, 229
correlation immune, 17
cryptanalysis, 157, 161
cryptographic protocol, 89, 235

data complexity, 173
DDoS attacks, 217
decisional Diffie-Hellman assumption, 187
denial-of-service attack, 151
differential factorization algorithm, 25
differential-linear cryptanalysis, 173
digital signature, 67, 107, 113, 123, 161
digital signature scheme, 97
discrepancy transform, 1
discrete logarithm problem, 67, 73
discrete logarithms, 107

electronic cash system, 73
ElGamal cryptosystem, 81, 137
encoding efficiency, 229
encryption, 165
equation solving, 145
EV-DO, 181

factoring, 107
Fiat-Shamir identification scheme, 129
filtering generator, 1

GOST, 123
group signature, 89, 129

ID-based cryptography, 59
identity based signature, 97
improved RSA cryptosystem, 137
IP traceback, 217

key escrow, 137

Legendre sequence, 9
linear codes, 33
linear complexity, 9

message recovery, 81
model checking, 235
modified Jacobi sequence, 9
modified polyphase Jacobi sequence, 9
multi-signature, 157

network security, 209
new criterion for secure RSA moduli, 25
nonlinearity, 17

one-off blind public key, 129

P2P network, 209
packet marking, 217
pairing, 165
periodic autocorrelation functions, 9
permutations, 1
privacy and anonymity, 89
provable security, 81
proxy signature, 53, 59, 89, 161
proxy signcryption, 53

QP, 229
quadratic residue, 97
quantum computation, 195
quantum cryptographic algorithm, 195
quantum cryptography, 201
quantum key distribution, 201
quantum relay, 201

resilient function, 17
robustness, 137
running-mode analysis, 235

secret sharing scheme, 81
secure multi–party computation, 145
security analysis, 181
signcryption, 53
standard complexity model, 187

stratospheric platform, 201

threshold group-signature scheme, 81
threshold scheme, 137
time complexity, 173
TMN protocol, 235
transitive digital signature, 113

Zhu-Lee-Deng's scheme, 187